SCHOLASTIC

Soda Bottle Science

25 Easy, Hands-on Activities That Teach Key Concepts in Physical, Earth, and Life Sciences—and Meet the Science Standards

by Steve "The Dirtmeister®" Tomecek

New York • Toronto • London • Auckland • Sydney
New Delhi • Mexico City • Hong Kong • Buenos Aires

Teaching Resources

Dedication

To my good friend John Farina,
who never let an empty soda
bottle go to waste!

Cover and interior design by NEO Grafika Studio
Illustrations by Mike Moran

ISBN-13: 978-0-439-75465-1
ISBN-10: 0-439-75465-8

5 6 7 8 9 10   40   15 14 13 12 11 10 09 08

Table of Contents

Introduction 5

National Science Education Standards 7

PHYSICAL SCIENCE

Cartesian Diver 10

Make a model submarine to explore how buoyancy can be changed.

Wave Machine 12

Investigate how differences in density can be used to test a physical change in matter.

Density Dilemma 14

Discover how the density of water changes when it turns from solid to liquid.

Solutions, Suspensions, and Mixtures 16

Explore the differences between suspensions and solutions.

Soda Bottle Thermometer 18

Discover the relationship between temperature and density in a liquid while building a model thermometer.

Viscosity Bottles 20

Test how viscosity differences in different liquids help control their flow.

Sonic Bottles 22

Use different-sized soda bottles to change the pitch of sound.

Soda Bottle Magnifier 24

Create a simple magnifier using a soda bottle as a lens.

Inertia Bottle 26

Test Newton's first law of motion with a soda bottle and a ramp.

Soda Bottle Pendulum 28

Discover the "law of the pendulum" using a soda bottle and string.

EARTH SCIENCE

Water Cycle in a Bottle 31

Use a soda bottle to make a model of the water cycle.

Hot-Air Balloon 33

Find out what makes the wind blow and investigate how a change in air temperature affects air's density.

Whirlpool in a Bottle 35

Discover how whirlpools and hurricanes result from the dynamics of fluid flow.

Rain Gauge 37

Build a simple rain gauge using different-sized soda bottles.

Infiltration Bottles 39

Explore how different soil types help control the flow of ground water.

Porosity Bottle 41

Discover which type of soil is best at retaining moisture.

Sediment-Settling Bottles 43

Use soda bottles to help separate different sediments found in soil.

Water-Pressure Bottle 45

Discover how a change in depth creates a change in water pressure.

LIFE SCIENCE

Soda Bottle Terrarium 47

Use a soda bottle to make a model ecosystem.

Soda Bottle Bug Habitat 50

Monitor the health and growth of isopods using a soda-bottle bug habitat.

Soda Bottle Greenhouse 52

Test and observe how the color of light impacts the growth of plants.

Transpiration Bottle 55

Investigate to see under what conditions plants lose water.

Colored Leaves 57

Discover how plants transport water through their tissue.

Yeast Beast 59

Find out what conditions are best for the growth of yeast.

Soda Bottle Composters 61

Use soda bottles to discover how natural nutrient recycling works.

Introduction

No question about it—the most effective science lessons have some type of hands-on component to them! Let's face it, when students get to "play," not only does their level of interest go up, but they often retain more information and have a better understanding of the science concepts being presented. Unfortunately, gathering enough materials for a full class of students to conduct a hands-on science lesson can sometimes be cost prohibitive. Between the high cost of science kits and the lead time needed to order them, many teachers are forced to forgo hands-on activities and resort to simple "chalk-and-talk" lessons.

That's why we've developed this series of science activity books. Like its two companions, *Coffee Can Science* and *Sandwich Bag Science*, *Soda Bottle Science* features 25 standards-based science activities structured around the use of one or more empty plastic soda bottles. By following the simple instructions in this book and using a few other "dirt cheap" materials, you and your students will be able to conduct dozens of fun and easy hands-on experiments and projects related to the physical, earth, and life sciences. While some of the activities in this book have been adapted from previously published ideas, many more are original designs. They have all been developed and tested in actual classroom settings with elementary and middle-school students so you can be sure they all work! In the end, we hope that they'll not only inspire your students to learn more science, but also give you the inspiration to experiment and develop your own inexpensive science activities.

How to Use This Book

Each of the 25 activities in this book comes with its own teaching guide and reproducible lab sheet for student use. The teaching guide features background information about the key science concepts in the activity, plus a mini-lesson and demonstration to introduce students to the activity. The reproducible page offers easy step-by-step instructions on how to conduct the activity, plus critical-thinking questions that invite students to make and write their own predictions, record results, and explain the outcomes of their experiment—a great way to combine science and writing!

There are several ways you can use or present the activities in this book:

◆ Use the activities as teacher demonstrations to spark students' interest and get them revved up for an upcoming lesson on a related science topic.

◆ Have students conduct the hands-on activities as directed. To make the activity more manageable, you may want to divide the class into small groups where students work together to do the activity.

◆ Set up "stations" around the classroom and present related activities at the same time. For example, you can present Wave Machine (page 12), Soda Bottle Thermometer (page 18), and Hot-Air Balloon (page 33) together and have students investigate the density of matter. When you set up multiple stations around the room, students can rotate to each one and see how they all relate to the same concept.

Whichever option you choose be sure to provide each student with his or her own copy of the lab sheet. This way, each student understands what the activity is about and can record his or her own predictions and observations.

Collecting Soda Bottles

It goes without saying that you will need many soda bottles to complete the activities in this book. Ask students to bring in clean, empty soda bottles with screw caps from home. Several experiments use both 1-liter and 2-liter bottles. Most of the activities require the bottles to be free of labels and the plastic rings that are often left behind on the neck when the cap is unscrewed. Both of these can easily be removed with a pair of sharp scissors. For safety's sake, you might want to do this yourself! Many of the activities require you to cut the bottles to a specific shape or size before they are used. Again, this can be done with a good pair of adult scissors that have a point on at least one end. When cutting the bottles, be careful not to slip and injure yourself!

National Science Education Standards

The activities in this book meet the following science standards:

Physical Science

For Grades K–4:
Properties of Objects and Materials
- ✦ Objects have many observable properties, including temperature. Those properties can be measured using tools, such as thermometers.
- ✦ Objects can be described by the properties of the materials from which they are made.
- ✦ Materials can exist in different states—solid, liquid, and gas. Some materials can be changed from one state to another by heating or cooling.

Position and Motion of Objects
- ✦ The position of an object can be described by locating it relative to another object or background.
- ✦ The position and motion of objects can be changed by pushing and pulling. The size of the change is related to the strength of the push or pull.
- ✦ Sound is produced by vibrating objects. The pitch of a sound can be varied by changing the rate of vibration.

Light, Heat, Electricity, and Magnetism
- ✦ Light travels in a straight line until it strikes an object. Light can be reflected, refracted, or absorbed.
- ✦ Heat can be produced in many ways. Heat can move from one object to another.

For Grades 5–8:
Properties and Changes of Properties in Matter
- ✦ A substance has characteristic properties such as density and solubility. A mixture of substances often can be separated into the original substances.

Motions and Forces
- ✦ An object that is not being subjected to a force will either stay at rest or continue to move at a constant speed and in a straight line.
- ✦ Unbalanced forces will cause changes in the speed or direction of an object's motion.

Transfer of Energy

✦ Energy is a property of many substances and is associated with heat, mechanical motion, sound, and light. Energy is transferred in many ways.

✦ Heat moves in predictable ways, flowing from warmer objects to cooler ones.

✦ Light interacts with matter by transmission and absorption.

✦ A tiny fraction of the light from the sun reaches the earth, transferring energy from the sun to the earth.

Earth and Space Science

For Grades K–4:
Properties of Earth Materials

✦ Earth materials are solid rocks and soils, water, and gases of the atmosphere. The varied materials have different physical and chemical properties.

✦ Soils have properties of color, texture, and the capacity to retain water.

Changes in the Earth and Sky

✦ The surface of the earth changes. Some changes are due to slow processes such as erosion and weathering.

✦ Weather changes from day to day. Weather can be described by measurable quantities, such as temperature and precipitation.

For Grades 5–8:
Structure of the Earth System

✦ Soils are often found in layers, with each having a different composition and texture.

✦ Some changes in the earth can be described as the "rock cycle." Old rocks at Earth's surface weather, forming sediments that are deposited and buried.

✦ Water, which covers the majority of the earth's surface, circulates through the "water cycle." Water evaporates from the earth's surface, rises and cools as it moves to higher elevations, condenses, and falls back to the surface.

Life Science

For Grades K–4:
Characteristics of Organisms

✦ Organisms have basic needs. Animals require air, water and food; plants require air, water, nutrients, and light. Organisms can survive only in environments where their needs are met.

✦ Each plant or animal has different structures that serve different functions in growth, survival, and reproduction.

Life Cycles of Organisms

✦ Plants and animals have life cycles that include being born, developing into adults, reproducing, and dying.

Organisms and Their Environments

✦ All animals depend on plants. Some animals eat plants for food, and other animals eat animals that eat plants.

✦ All organisms cause changes in the environment where they live.

For Grades 5–8:
Populations and Ecosystems

✦ A population consists of all individuals of a species that occur together in a given place and time. All populations living together and the physical factors with which they interact compose an ecosystem.

✦ Populations or organisms can be categorized by the function they serve in an ecosystem.

✦ For ecosystems, the major source of energy is sunlight. Energy entering the ecosystem as sunlight is transferred by producers into chemical energy through photosynthesis.

✦ The number of organisms an ecosystem can support depends on the resources available and abiotic factors.

Science and Technology

✦ People have invented tools and techniques to solve problems.

✦ Tools help scientists make better observations and measurements.

Cartesian Diver

Get It Together

- 1-liter soda bottles with caps (for each group of students)
- clear 2-liter bottle (for demo)
- 2 6-inch balloons
- water
- "Diver Down" lab sheet (page 11)

Fizzy Science

Buoyancy is the ability of an object to float. Some objects float because they are less dense than the surrounding water. *Density* is the mass of an object divided by its volume (D=M/V). Water has a density of about 1 gram per cubic centimeter. If an object's density is less than this, it will be positively buoyant and float on water. If its density is greater, it will be negatively buoyant and sink. Submarines control their total density using *ballast tanks*, which can be filled with either air or water. When its tanks are filled with air (which is less dense than water), a sub rises. When they're filled with water, the sub sinks. The Cartesian diver in this activity works the same way—when you squeeze the bottle, the air in the dropper gets compressed and the dropper fills with water, making the diver sink. Releasing the pressure on the bottle causes the air inside the dropper to expand, making the dropper lighter and causing it to float.

Before You Start

Remove the labels from all soda bottles. Cut the top off the 2-liter bottle to make a cylinder and fill it ¾ of the way with water. Make a sample Cartesian diver (see page 11).

What to Do

❶ Set up the 2-liter cylinder with water in front of the class. Explain that they will be investigating buoyancy. Ask: What is buoyancy? *(The ability of an object to float)*

❷ Fill one balloon with water and knot it closed. (Make sure it fits in the cylinder.) Blow up the other balloon until it's about the same size as the first one and knot it closed.

❸ Hold up both balloons and ask the class: What will happen when I put the air-filled balloon in the water? *(It will float on the water.)* Put the air-filled balloon in the water to demonstrate. Next, have the class predict what will happen when you place the water-filled balloon in the water. *(It will sink below the water.)* Demonstrate.

❹ Ask: Why did one balloon float and the other sink? *(Some students may say that the air-filled balloon is lighter than the water-filled one.)* Explain that the reason is actually *density*—how much mass something has compared to how much space it takes up. Water is denser than air, so the water-filled balloon is less buoyant than the air-filled balloon.

❺ Hold up the sample Cartesian diver. Gently squeeze the bottle to make the diver sink and then release it again so that it rises. Explain that, like this diver, submarines can control its buoyancy by using air and water. Hand out copies of the "Diver Down" lab sheet and invite students to find out how.

Diver Down

How does a submarine change its buoyancy?

❶ Fill the soda bottle so that the water is about $2\frac{1}{2}$ centimeters (1 in.) from the top. Fill the small cup with water. Fill the glass medicine dropper about halfway with water.

❷ Place the dropper into the bottle of water and screw the cap on tight. The dropper should be floating on top of the bottle. If it sinks, empty the bottle, take the dropper out, and try it again with less water in the dropper.

❸ Look closely at the dropper in the bottle. How much of the dropper is filled with water and how much of it is filled with air?

_____

_____

You'll Need

- 1- or 2-liter soda bottle with cap and the label removed
- glass medicine dropper (available at most pharmacies)
- small plastic cup
- water
- ruler

❹ Grasp the outside of the soda bottle firmly with two hands and squeeze it tight until the dropper sinks to the bottom of the bottle. Why do you think this happens?

_____

_____

❺ With the dropper still at the bottom of the bottle, look closely at the water level inside the dropper. How has it changed? Why do you think this happens?

_____

_____

❻ Predict what will happen when you release the pressure on the outside of the bottle. Write your prediction here:

_____

_____

❼ Release the pressure on the bottle and observe what happens. How do you think you can make the dropper stay in the middle of the bottle? Try it.

Think About It: Ballast tanks in a submarine can be filled with either water or air. How do you think they make a submarine sink or float?

Wave Machine

Get It Together

- 2 clear 1-liter soda bottles (for each group of students)
- clear 2-liter bottle (for demo)
- sharp scissors
- ruler
- small rock
- wooden block, about the same size as the rock
- water
- "Catch a Wave" lab sheet (page 13)

Fizzy Science

When it comes to properties of matter, a difficult concept for some people to grasp is *density*. Density is a measure of how much mass an object has compared to its volume. It can be expressed in the equation Density = Mass / Volume (D=M/V), and is measured in grams per cubic centimeter (g/cc). Things that have high density may feel heavy, while objects with low density may feel light. Most rocks and metals have a relatively high density, while things like feathers and foam rubber have very low densities. Density is also a property of liquids and gases. A good example is water and oil. Freshwater has a density of 1 g/cc, and it is the standard around which the metric system is built. Most oils have a density less than 1 g/cc, so liquid oil generally will float on liquid water. By using water as a standard, scientists can quickly compare the density of different liquids and, in the process, create a really cool toy!

Before You Start

Remove the labels from all soda bottles. Cut all the tops off the bottles 15 centimeters (6 in.) from the bottom to make cylinders. Fill the 2-liter-bottle cylinder ¾ of the way with water.

What to Do

❶ Explain to students that they will be investigating density. Ask: What's density? *(A measure of how much mass something has compared to the amount of space it takes up)*

❷ Pass around the wooden block and the rock and ask students to compare their weights. Explain that the rock feels heavier because it has a greater density than the block. In other words, it has more matter packed into the same amount of space.

❸ Place the 2-liter-bottle cylinder with water where everyone can see. Ask: What do you think will happen when I place the block and the rock in the water? *(The rock will sink and the block will float.)* Call on a student volunteer to test the class's predictions.

❹ Explain that the wooden block floats because it is less dense than water, while the rock has a greater density than that of water.

❺ Ask: Are solids the only types of matter that have different densities? Encourage students to express their opinions as you hand out copies of the "Catch a Wave" lab sheet.

Catch a Wave

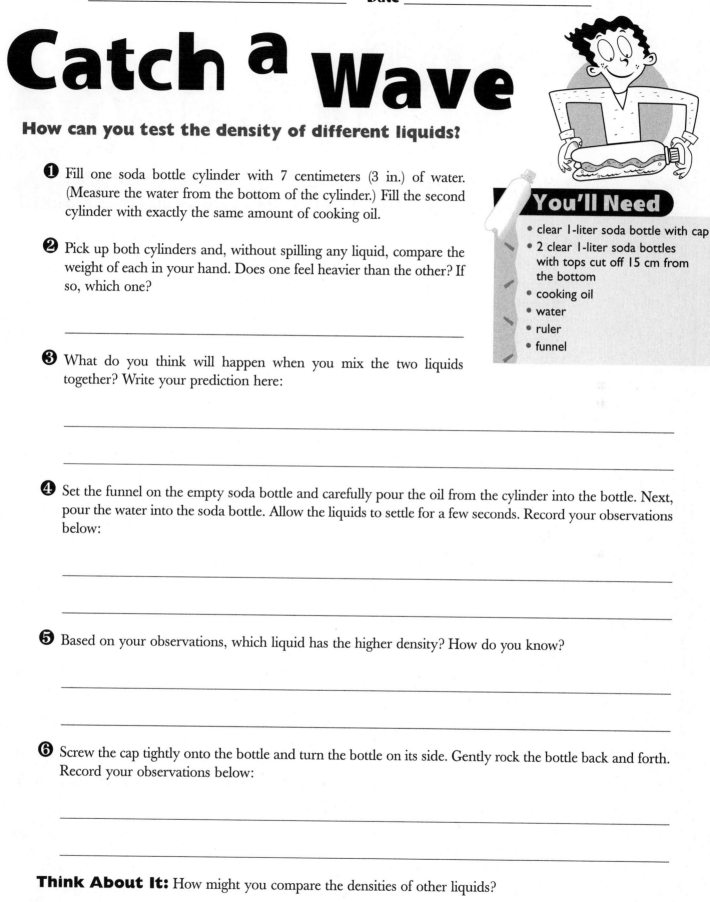

How can you test the density of different liquids?

❶ Fill one soda bottle cylinder with 7 centimeters (3 in.) of water. (Measure the water from the bottom of the cylinder.) Fill the second cylinder with exactly the same amount of cooking oil.

❷ Pick up both cylinders and, without spilling any liquid, compare the weight of each in your hand. Does one feel heavier than the other? If so, which one?

_____

❸ What do you think will happen when you mix the two liquids together? Write your prediction here:

_____

_____

❹ Set the funnel on the empty soda bottle and carefully pour the oil from the cylinder into the bottle. Next, pour the water into the soda bottle. Allow the liquids to settle for a few seconds. Record your observations below:

_____

_____

❺ Based on your observations, which liquid has the higher density? How do you know?

_____

_____

❻ Screw the cap tightly onto the bottle and turn the bottle on its side. Gently rock the bottle back and forth. Record your observations below:

_____

_____

Think About It: How might you compare the densities of other liquids?

You'll Need

- clear 1-liter soda bottle with cap
- 2 clear 1-liter soda bottles with tops cut off 15 cm from the bottom
- cooking oil
- water
- ruler
- funnel

Density Dilemma

Get It Together

- 3 clear 1-liter soda bottles (for each group of students)
- 2 clear 1-liter soda bottles (for demo)
- sharp scissors
- ruler
- water
- bottle of corn oil
- stick of 100% corn-oil margarine
- 2 ice cubes
- plastic knife
- "Ice Is Nice" lab sheet (page 15)

Fizzy Science

Water is an unusual substance. When most substances change from liquid to solid, they become more compact and, as a result, their density increases. When liquid water freezes, its volume increases. Since its mass remains the same, as frozen water expands, its density decreases. As a result, ice floats on water. If water behaved like most substances, ponds would freeze in winter from the bottom up instead of the top down, killing all the living things in them. In addition, instead of having ice caps floating near the poles, oceans would freeze solid, cooling the entire earth and putting us in a permanent ice age.

Before You Start

Remove the labels from all soda bottles. Cut all the tops off the bottles, about 15 centimeters (6 in.) from the bottom, to make cylinders. Make sure the ice cubes for the activity stay frozen. If they begin to melt, their density will increase, and they will sink in the oil.

What to Do

❶ Tell students that they are going to investigate how the density of water changes when it changes from one state of matter to another. Place two soda-bottle cylinders in front of the room. Fill one cylinder halfway with water and the other halfway with corn oil. Ask: What state are the water and corn oil in? *(Liquid)*

❷ Use the knife to cut off a $2\frac{1}{2}$-centimeter (1-in.) piece of margarine. Hold up the margarine, explaining that it is a solid piece of corn oil. Have students predict what will happen to the margarine when you place it in the liquid corn oil. *(It will sink.)* Ask: Is the margarine more dense or less dense than the oil? *(More dense)*

❸ Next, hold up an ice cube, reminding students that it is frozen water. Ask students to predict what will happen if you drop the ice in the water. *(It will float.)* Explain that unlike most substances, water becomes less dense when it changes from liquid to solid. Most common substances, like margarine, become denser when they turn solid and sink in their liquid form.

❹ Ask: What do you think will happen if you drop ice into liquid oil? Hand out copies of the "Ice Is Nice" lab sheet and invite students to find out.

Ice Is Nice

What happens to the density of water when it changes from a solid to a liquid?

❶ Fill two cylinders with 7 centimeters (3 in.) of oil. (Measure from the bottom of the cylinder.) Fill the third cylinder with exactly the same amount of water.

❷ Prepare to drop an ice cube in the cylinder with the water. Will the ice sink or float? Write your prediction here:

_____

❸ Place the ice in the water and observe what happens. Was your prediction correct? What does this tell you about the density of ice compared to the density of water?

_____

_____

❹ Fill the small cup with water and prepare to pour it into one of the cylinders of oil. Predict: What will the water do—sink or float?

_____

❺ Pour the water into the oil and observe what happens. Was your prediction correct? What does this tell you about the density of water compared to the density of oil?

_____

_____

❻ Based on your first two trials, predict what will happen if you place a piece of ice in the second cylinder of oil. What will happen when the ice begins to melt and turn into liquid water?

_____

❼ Place one or two ice cubes in the second cylinder of oil and observe what happens when the ice starts to melt. Record your observations here:

_____

_____

Think About It: What can you conclude about the density of water when it changes from solid to liquid?

You'll Need

- 3 clear 1-liter soda bottles with tops cut off 15 cm from the bottom
- cooking oil
- water
- 3 ice cubes
- ruler
- 6-oz plastic cup

Solutions, Suspensions, and Mixtures

Get It Together

- 3 clear 2-liter soda bottles (for each group of students)
- 3 large (12–16 oz) clear plastic cups, ¾ full with water
- plastic teaspoon
- salt
- powdered milk
- cup of sand
- "In the Mix" lab sheet (page 17)

Fizzy Science

We live in a mixed-up world! Most substances we encounter are mixtures of different materials. Mixtures come in two forms. In *heterogeneous mixtures,* you can physically see the different components; for example, sand in water. In *homogeneous mixtures,* the different components are evenly spread so that they have a uniform composition throughout. Homogeneous mixtures can be either *solutions* or *suspensions.* In a solution, solid particles are completely dissolved so that the entire solution is in a liquid phase, like salt water. In suspensions, solid particles are spread evenly but remain solid, floating in the liquid. Milk and blood are good examples. Given enough time, the different components in a suspension will separate under the force of gravity.

Before You Start

Remove the labels from all soda bottles.

What to Do

❶ Tell students that they're about to get all mixed up as they experiment with mixtures. Ask: What is a mixture? *(A substance made from two or more materials mixed together)*

❷ Ask a student volunteer to assist you. Have the student stir a teaspoonful of sand into the cup of water. Ask the class to observe what happens as soon as the volunteer stops stirring. (The sand and water immediately separate.) Explain that sand and water make a *heterogeneous mixture* because they don't stay mixed. They separate.

❸ Call on another volunteer to stir a teaspoonful of powdered milk into another cup of water. Ask the class to compare this with the sand–water mixture. (The milk mixture is cloudy but the particles are spread evenly in the water.) Explain that the milk–water mixture is called a *suspension* because the little milk particles are floating evenly throughout the water. Over time, however, they will also begin to separate.

❹ Invite a third volunteer to stir a teaspoonful of salt into the third cup of water. Have the class compare the salt water to the first two mixtures. (The water may be a little hazy at first but the salt will disappear into the water.) Explain that the saltwater mixture is called a *solution.* You cannot separate the parts of a solution through physical means.

❺ Distribute copies of the "In the Mix" lab sheet and invite students to conduct their own solution/suspension tests.

In the Mix

How are suspensions different from solutions?

You'll Need

- 3 clear 2-liter soda bottles with caps
- water
- measuring cup
- 1 tsp of baking soda
- 1 tsp of cornstarch
- 1 tsp of sugar
- 3 pieces of 3-by-5-inch paper

❶ Fill each soda bottle with about 16 ounces of water.

❷ Set the three pieces of paper on the table in front of you. On the first, put a teaspoonful of sugar; on the next, a teaspoonful of cornstarch; and on the last, a teaspoonful of baking soda.

❸ Carefully observe each powder. Pick a tiny pinch of each one and feel the texture in your fingers. Write down as many properties of each powder as you can below:

Sugar: _____

Cornstarch: _____

Baking soda: _____

❹ Based on each powder's properties, predict whether each substance will make a solution or a suspension when it mixes with the water. Write your predictions here:

Sugar: _____

Cornstarch: _____

Baking soda: _____

❺ Carefully lift each paper and pour each powder into a different bottle of water. Place the caps on the three bottles and shake each vigorously for 10 seconds. Allow the bottles to stand for about one minute and then observe the mixtures. Record your observations for each below:

Sugar: _____

Cornstarch: _____

Baking soda: _____

❻ Which substances formed solutions and which ones formed suspensions? How can you tell?

_____

_____

Think About It: How could you use this system to test other mixtures?

Soda Bottle Thermometer

Get It Together

- 1-liter soda bottle with cap (for each group of students)
- electric drill
- liquid-filled lab thermometer (lab grade or household)
- plastic bowl with ice
- plastic bowl with warm water
- "The Heat Is On" lab sheet (page 19)

Fizzy Science

Thermometers help you decide what to wear outdoors and alert you when your body is out of whack! The basic liquid-filled thermometer works through *thermal expansion*. When most liquids get hot, their molecules begin to vibrate faster and spread apart, causing the overall volume of the liquid to increase. When liquids cool, their molecular energy decreases and they begin to contract. By trapping a liquid in a tube and putting a simple scale behind it, you can measure the change in thermal energy or temperature by watching the liquid move up and down. Most modern thermometers use colored alcohol. The very first thermometers used water, but because water's rate of expansion is fairly small, these thermometers had to be very large.

Before You Start

Remove the labels from all soda bottles. Use an electric drill to make a ¼-inch hole through the top of each bottle cap.

What to Do

❶ Ask students: How can we tell what the temperature is? *(Read a thermometer)* Explain that thermometers have been around for hundreds of years. While there have been many changes in their design, the basic idea behind how a thermometer works hasn't changed much.

❷ Hold up a thermometer, walking around the room so students can see it up close. Ask them to describe it. *(It has a glass tube filled with liquid and has a scale.)*

❸ Invite a student volunteer to read the temperature on the thermometer, then place it into the bowl of ice. Ask students: What will happen to the liquid in the tube? *(It will go down.)* After a minute, remove the thermometer from the ice and have another volunteer read the temperature to confirm students' prediction.

❹ Next, place the thermometer in the bowl of warm water and have students predict what will happen to the liquid. *(It will rise.)* Explain that the liquid goes up and down due to *thermal expansion*. When matter gets hot it expands, and when it gets cold it contracts.

❺ Tell students that they will build their own water-filled thermometers. Hand out copies of "The Heat Is On" lab sheet.

The Heat Is On

How does a thermometer work?

❶ Place a few drops of red food coloring into the empty soda bottle and slowly fill it to the top with cool tap water. As you fill the bottle, gently swish it around so that the color mixes evenly. Tightly screw on the cap with the hole.

❷ Wrap a piece of tape around the straw about 8 centimeters (3 in.) from one end. With the taped side pointing down, slip the straw into the hole in the cap so that the bottom of the straw extends into the water. The tape should keep the straw from slipping through the hole. Some water will move up into the straw and may spill out, so be prepared with a paper towel!

You'll Need

- 1-liter soda bottle
- bottle cap with a ¼-inch hole drilled through the top
- red food coloring
- white glue
- large, clear plastic drinking straw
- cool water
- ruler
- tape
- ice
- warm water
- large pan or bowl
- paper towel (for cleanup)

❸ Use white glue to seal the space around the opening where the straw enters the cap. This must be an airtight seal for the thermometer to work properly! Allow the glue to dry for about 30 minutes.

❹ After the glue has dried, pour some water out of the bottle so that only about 2½ centimeter (1 in.) of water is visible above the cap in the straw. You are now ready to test your thermometer.

❺ Place your water thermometer into the large empty bowl and start filling the bowl with warm water. Predict: What will happen to the water in the straw?

_____

_____

❻ Observe the straw. Does your observation match your prediction? Next, empty the bowl and repeat step 5, this time packing several ice cubes around the base of the bottle. Predict: What will happen this time?

_____

_____

❼ Based on your observations, what might happen to the water in the straw if the outside temperature got too hot?

_____

_____

❽ Would your thermometer work if the temperature outside fell below zero? Explain why.

_____

_____

Think About It: How might you use a real thermometer to make a scale that would work on your soda bottle thermometer?

Viscosity Bottles

Get It Together

- 2 clear 1-liter soda bottles (for each group of students)
- 2 large (16-oz) plastic cups
- "Go With the Flow" lab sheet (page 21)
- 8-oz clear plastic cup of molasses or dark corn syrup
- 8-oz clear plastic cup of water, dyed brown with food coloring

Fizzy Science

All liquids are fluids—they have no definite shape and they flow from one place to another—but they don't all flow the same way. Some liquids are very thick and gooey and flow very slowly, while other liquids are relatively thin and flow quickly. The property that measures the resistance to fluid flow is called *viscosity*. A liquid's viscosity is controlled by the internal friction between individual molecules. The greater the internal friction, the "stickier" a liquid is and the higher its viscosity will be. Liquids like pancake syrup and molasses have high viscosity, while water, soda, and alcohol have a relatively low viscosity. Cooling a liquid generally increases its viscosity while heating it decreases its viscosity. In some liquids, such as water, temperature has little effect on viscosity. It stays very fluid until it freezes!

Before You Start

Remove the labels from all soda bottles.

What to Do

❶ Tell students that they're going to watch a race . . . not between people or cars, but between liquids! Call on two student volunteers. Set up two large cups in front of the room. Give one volunteer the cup of water and the other the cup of syrup.

❷ Have the volunteers hold up each cup for the class to see, but don't tell the class what the two liquids are. Have students compare the properties of the two liquids and predict which will flow faster. On the count of three, have the volunteers empty their cups without spilling. (The water will flow instantly while the syrup will take longer.)

❸ Ask students if they have ever heard the expression "slow as molasses." Well, now they know why! Inform students that the liquid that flowed slowly was molasses, which is like syrup, while the other one was water.

❹ Explain that the two liquids flowed at different speeds due to a property called *viscosity*. Viscosity has to do with the amount of friction between the individual molecules in a liquid. The higher the viscosity, the greater the friction and the slower the liquid flows.

❺ Tell students that they will investigate how temperature affects viscosity. Hand out copies of the "Go With the Flow" lab sheet.

Go With the Flow

Does the viscosity of a liquid change with temperature?

❶ Fill one bottle with 24 ounces of cooking oil and the second with an equal amount of water. Screw the caps on both bottles tightly and place them in a freezer or cooler filled with ice for 15 minutes before starting the experiment.

❷ After the liquids have been chilled, begin your viscosity test. Open the caps to allow air to get into the bottles so that they expand to full size and then close the caps tightly again. Hold the two bottles side-by-side. Tilt the bottles gently and observe how the liquid in each behaves. Record your observations:

Water:_____

Oil: _____

❸ Open each bottle. Prepare to drop a penny into each bottle to compare the viscosity of the two liquids. The faster the penny falls, the lower the viscosity. Based on your observations of the two liquids, which one do you think will have a higher viscosity? Why?

_____

❹ Drop a penny into the bottle of water and observe how fast it falls. Now do the same with the bottle of oil. Record your observations below:

_____

❺ Next, you will test to see if changing the temperature of a liquid affects its viscosity. Allow each bottle to sit out in the warm air for 15 minutes or use a hair dryer to heat them up for about 5 minutes each. Repeat step 4 using the warm liquids. Record your observations below:

_____

_____

❻ Based on your observations, do you think that temperature affects the viscosity of all liquids in the same way? Why?

_____

_____

Think About It: Why do you think it's not a good idea to keep pancake syrup or honey in the refrigerator?

You'll Need

- 2 clear 1-liter soda bottles
- 24 oz of cooking oil
- 24 oz of water
- 4 pennies
- electric hair dryer
- freezer or cooler filled with ice
- watch or clock

Sonic Bottles

Get It Together
- 1-liter soda bottle (for each group of students)
- 2-liter soda bottle (for each group of students)
- empty $\frac{1}{2}$-liter water bottle
- "Pitch Pipes" lab sheet (page 23)

Fizzy Science

One of the first musical instruments to be developed was the pitch pipe. A simple pitch pipe is nothing more than a hollow tube originally made out of dried reeds. By cutting tubes to different lengths, people found that they could tune the various pipes to different musical notes. Different-sized tubes produce different notes because they contain different volumes of air. Sound happens when air vibrates back and forth. In general, the greater the volume of air that's vibrating, the deeper the note. This can be seen in most wind instruments today. A piccolo produces high-pitched notes, while a tuba produces only low tones. When a musician plays different notes on most wind instruments, she usually opens and closes holes along the length of the instrument, effectively shortening and lengthening the size of the column of air vibrating inside.

Before You Start

Remove the labels from all soda bottles.

What to Do

❶ Tell students that they're in for a special treat because you're going to entertain them with a little music. Pick up an empty 2-liter soda bottle and blow across the top so that you make a steady tone.

❷ Next, pick up the 1-liter bottle. Hold it up next to the 2-liter bottle and ask students: How will this bottle sound compared to the first? *(The 1-liter bottle will produce a higher pitch.)* Blow across the top of the 1-liter bottle and compare the sound to the 2-liter bottle. It will indeed sound higher.

❸ Next, pick up the empty $\frac{1}{2}$-liter water bottle. Before you blow into it, ask students to predict how it will sound compared to the first two. *(It will have the highest pitch.)*

❹ After testing it to confirm their prediction, explain that each bottle makes a sound because of the air inside. When you blow across the top, the air begins to vibrate, and when something starts vibrating, it makes a sound.

❺ Pose this question: Since all three bottles have air in them, why does each one produce a different sound? Is it because they have different widths, different lengths, or different volumes? Challenge students to discover the sound-sational answer as you hand out copies of the "Pitch Pipes" lab sheet.

Pitch Pipes

What factor controls the pitch of a wind instrument?

❶ Blow across the top of the 1-liter soda bottle and listen to the pitch of sound it makes. Now blow across the top of the 2-liter bottle. How do the two sounds compare? Why do you think they sound this way? Write your ideas below:

_____

_____

You'll Need
- 1-liter soda bottle
- 2-liter soda bottle
- water
- funnel
- ruler

❷ Take the 2-liter soda bottle and, using the funnel, add about 5 centimeters (2 in.) of water to the bottle. Try blowing across the bottle again. How did the sound change? Why do you think that is?

_____

❸ Continue filling the 2-liter soda bottle with water until it has the exact same pitch as the 1-liter bottle. Compare them frequently by blowing across each bottle. When you get the two bottles sounding exactly the same, stop adding water and observe the two bottles. How are they similar? How are they different?

_____

❹ Based on your observations, which factor do you think is responsible for the two bottles having the same pitch—the shape of the bottles or the volume of air in the bottles? Why do you think so?

_____

_____

❺ What volume of air do you think is left in the 2-liter bottle? How can you measure it?

_____

❻ Fill the 1-liter soda bottle to the top with water. Using the funnel, carefully pour the water from the 1-liter bottle into the 2-liter bottle. The amount of water you pour into the 2-liter bottle should equal the amount of air in the bottle. How does the volume of air remaining in the 2-liter bottle compare with the volume of air in the empty 1-liter bottle? Does this explain why the two bottles have exactly the same tone?

_____

_____

Think About It: What factor do you think is more important for determining the pitch of a wind instrument—its shape or the volume of air trapped in it?

Soda Bottle Magnifier

OBJECTIVE: To observe how a lens magnifies an image

Get It Together
- clear 1-liter soda bottle (for each group of students)
- magnifier
- flashlight
- "Eye Spy" lab sheet (page 25)

Fizzy Science

Look at a nearby object through a magnifier and it appears bigger. Look at a distant object and it appears upside down. Why? Magnifiers work using *refraction*. When light travels at an angle through a transparent material, the individual rays of light bend. Devices that bend light are called *lenses*, and they come in many shapes and sizes. A magnifying lens is a *biconvex lens*—its two sides curve out so that it's thickest in the middle. When light travels through a biconvex lens, the light rays on the outer edges bend toward the middle, causing all the rays to meet at one point, called the *focal point*. Light appears brightest at the focal point. Past this point, the light rays crisscross in space, flipping the image upside down. Convex lenses are found in telescopes, microscopes, eyeglasses, and in your own eye.

Before You Start

Remove the labels from all soda bottles. If necessary, soak the outside of the bottles in hot water to remove any remaining glue.

What to Do

❶ Walk around the room, looking through the magnifier and moving it back and forth in front of your eye. Ask students: What do you see happening? *(Your eye is getting bigger and smaller and occasionally your face may be upside down.)* What is this device that I'm using? *(A magnifier or magnifying glass)*

❷ Explain that a magnifier is also called a *biconvex lens* because it's thick in the middle and narrow along the edges on both sides. A magnifier makes things look bigger because it bends the light passing through it, bringing all the light rays in focus.

❸ Darken the room and stand about 3 meters (10 ft) in front of a wall. Aim the flashlight at the center. Have students observe the light spot as you slowly move away from the wall. Explain that as light travels from a source, as with a flashlight, its rays spread out. The farther away the source, the larger the area in which the light spreads and the dimmer it appears.

❹ Go back to your original spot and shine the light at the wall again, but this time, through the magnifier. Move the magnifier back and forth in front of the light and ask students to observe the spot. The spot on the wall will get bigger and smaller. Explain that by adjusting the position of the lens, you are changing the focal point of the light.

❺ Hand out copies of the "Eye Spy" lab sheet for students to make soda-bottle magnifiers.

Eye Spy

How does a lens focus light?

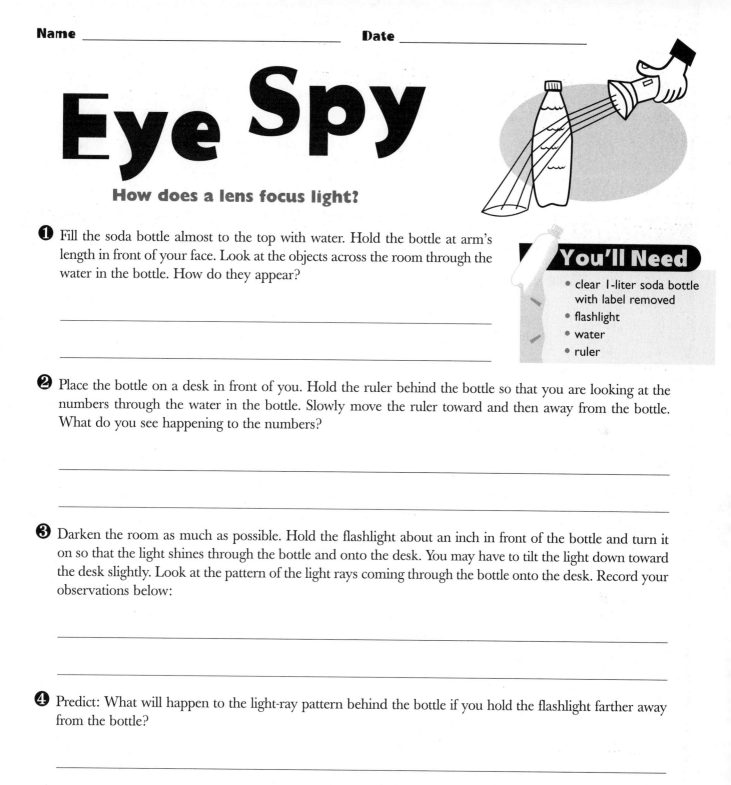

❶ Fill the soda bottle almost to the top with water. Hold the bottle at arm's length in front of your face. Look at the objects across the room through the water in the bottle. How do they appear?

_____

_____

You'll Need

- clear 1-liter soda bottle with label removed
- flashlight
- water
- ruler

❷ Place the bottle on a desk in front of you. Hold the ruler behind the bottle so that you are looking at the numbers through the water in the bottle. Slowly move the ruler toward and then away from the bottle. What do you see happening to the numbers?

_____

_____

❸ Darken the room as much as possible. Hold the flashlight about an inch in front of the bottle and turn it on so that the light shines through the bottle and onto the desk. You may have to tilt the light down toward the desk slightly. Look at the pattern of the light rays coming through the bottle onto the desk. Record your observations below:

_____

_____

❹ Predict: What will happen to the light-ray pattern behind the bottle if you hold the flashlight farther away from the bottle?

_____

_____

❺ Shine the light through the bottle again and watch the light-ray pattern on the desk as you begin to pull the light farther away from the bottle. How does this pattern on the desk explain your observations in steps 1 and 2?

_____

_____

Think About It: Do you think different-sized water bottles will magnify objects the same way? Experiment with 2- and 3-liter soda bottles to test your hypothesis.

Inertia Bottle

Get It Together
- sheet of newspaper
- 2-liter soda bottle with cap
- water
- paper towel
- "Bowling Bottle" lab sheet (page 27)

Fizzy Science

Heavy objects are generally much more difficult to move than light ones. While it's fairly easy to pull a child in a wagon, it would be difficult to push a car with the same child in it. The property of an object that describes the ease with which its motion can be changed is called *inertia*. Inertia is the resistance offered by an object to any change in its motion. The "law of inertia" is Newton's first law of motion, stating that a body at rest or in motion will stay at rest or in motion unless an outside force acts on it. Inertia is directly tied to the mass of an object. The more massive an object is, the more inertia it has.

What to Do

❶ Grab students' attention by telling them that you're going to perform a little magic trick. Fill a 2-liter soda bottle to the top with water and screw the cap on tightly. Dry the outside of the bottle with a paper towel. Spread a sheet of newspaper across a smooth surface (like a table or desk) in front of the class. Place the bottle on top of the sheet of paper. Grasp the edge of the paper and, with a sharp, quick tug, pull the newspaper out from under the bottle. The bottle of water should be left standing on the table.

❷ Ask: Why didn't the bottle move when I pulled the paper? *(You pulled the paper really fast and the bottle was fairly heavy.)*

❸ Ask students if they have ever seen a magician do this trick. If they have, ask them if they have ever seen it done with an object that wasn't fairly heavy. Most often, magicians will use heavy china plates, crystal goblets, or a stack of glasses that have been filled with liquid. Ask: What do you think would happen if we tried the trick using an empty soda bottle instead of a full one? *(The empty soda bottle would topple over.)* Empty the bottle and demonstrate.

❹ Explain that this trick works because of *inertia*. Inertia is the resistance offered by an object to any change in its state of motion. The more inertia an object has, the harder it is to change its motion. If something has a great deal of inertia and it's standing still, it will be hard to get it going. If something with a great deal of inertia is moving, it will be really hard to stop it.

❺ Hand out copies of the "Bowling Bottle" lab sheet and invite students to discover what property of matter controls inertia by doing a little bowling!

Bowling Bottle

How does the mass of an object affect its inertia?

❶ Clear a large open area on the floor. Create a ramp by laying the book flat on the floor and placing one end of the board on top of it. Hold the empty 2-liter bottle in one hand and the small ball in the other hand. Which feels heavier? Which do you think will have more inertia?

_____

You'll Need

- 2-liter soda bottle with cap
- large, flat wooden board (approximately 30-by-76-cm)
- thick book
- water
- small light ball, like a hand ball or tennis ball
- large heavy ball, like a basketball or soccer ball

❷ Place the bottle on the floor at the base of the ramp and set the small ball at the top of the ramp. Without pushing the ball, let it roll down the ramp and hit the bottle. What happens to the bottle when the ball strikes it? Why?

_____

❸ Take the bottle and fill it to the top with water. Screw on the cap tight. What have you done to the bottle's mass? What have you done to its inertia? Predict: What will happen to the bottle this time when the small ball rolls into it? Write your prediction here:

_____

❹ Repeat step 2 using the bottle full of water. What happened this time?

_____

_____

❺ Next, pick up the large ball. How much inertia does it have compared to the small ball? Predict what will happen to the full bottle of water when you roll the larger ball down the ramp. Why do you think so?

_____

_____

❻ Roll the large ball down the ramp and record your observations below. Based on your observations, how does the mass of an object affect its inertia and ability to move?

_____

_____

Think About It: Why do you think bowling pins and bowling balls must have a great deal of inertia?

Soda Bottle Pendulum

OBJECTIVE: To explore the law of pendulums

Get It Together
- 1-liter soda bottle with cap (for each group of students)
- electric drill
- thick string
- scissors
- ruler
- broomstick
- timer or watch with second hand
- "Swing Time" lab sheet (page 30)

Fizzy Science

When most people think about a pendulum, they visualize the type of device used to regulate a cuckoo or grandfather clock. But pendulums are all around us. In fact, you have a few pendulums hanging off your body! When you walk, your arms and legs work like pendulums, helping to keep you balanced. A pendulum is simply a mass that can swing freely back and forth from a fixed point.

Back in the late 1500s, Galileo investigated the motion of pendulums and discovered a few important rules about how they work. First, he noticed that the *period* of a pendulum (the time it takes to make one swing back and forth) is independent of the mass of the pendulum. A pendulum with a heavy weight at the bottom swings back and forth at the same rate as one with a light weight. He also discovered that the period was independent of the size of the arc of the swing. A pendulum that swung through a small arc did so at the same rate as a pendulum that swung through a large arc. The one factor that did impact the period of swing, however, was the pendulum's length. Given two pendulums with different lengths, the shorter one will swing back and forth faster.

Before You Start

Use an electric drill to make a ¼-inch hole through the top of each bottle cap. Precut 76-centimeter (30-in.) lengths of string for each group of students. Construct a model soda-bottle pendulum to share with students (see page 30). This activity is ideal for students working in small groups.

What to Do

❶ Ask students: What does a pendulum look like? *(A suspended weight that can swing freely back and forth)*

❷ Invite two student volunteers to be your "support group," asking them to hold each end of a broomstick. Tie the string attached to your soda bottle pendulum to the middle of the broomstick. The bottle should hang freely from the string.

❸ Pull back on the bottle to set the pendulum in motion. Have the class observe the way the pendulum moves. Explain that the time it takes for the pendulum to make one swing back and forth is called the *period*. By timing the periods, you can learn a great deal about pendulum motion.

(continued)

❹ Ask: Which do you think will have a longer period—a pendulum that swings through a big arc (one that has been pulled back a large distance) or one that swings through a small arc? Encourage students to share their ideas. Tell students that they can test their answers by counting how many times a pendulum swings back and forth in a specific amount of time.

❺ Ask a student volunteer to serve as the "timer." The class will count how many times (how many periods) the pendulum swings in 10 seconds. Start with a small arc. Pull the pendulum back from the center point only a short distance. Have the timer say "go" and, after 10 seconds, say "stop." Record the number of swings the pendulum made.

❻ Repeat step 5 with a big arc. You should get approximately the same number of swings for the pendulum. Explain that as long as the length and mass of the pendulum stays the same, the period of a pendulum will be the same even if the distance it swings is changed.

❼ Hand out copies of the "Swing Time" lab sheet and invite students to investigate how changing the length of the pendulum affects its period of motion.

Swing Time

How does the length of a pendulum affect its swing rate?

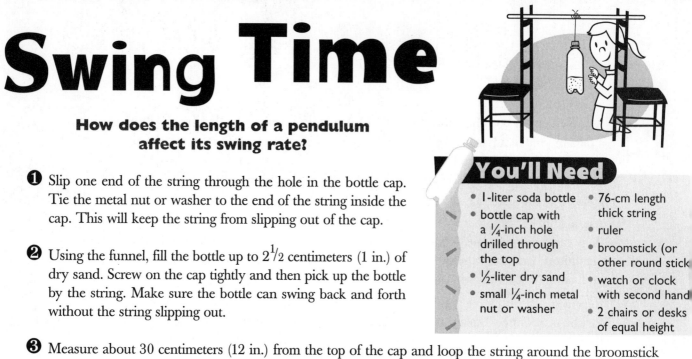

❶ Slip one end of the string through the hole in the bottle cap. Tie the metal nut or washer to the end of the string inside the cap. This will keep the string from slipping out of the cap.

❷ Using the funnel, fill the bottle up to $2\frac{1}{2}$ centimeters (1 in.) of dry sand. Screw on the cap tightly and then pick up the bottle by the string. Make sure the bottle can swing back and forth without the string slipping out.

You'll Need

- 1-liter soda bottle
- bottle cap with a ¼-inch hole drilled through the top
- ½-liter dry sand
- small ¼-inch metal nut or washer
- 76-cm length thick string
- ruler
- broomstick (or other round stick
- watch or clock with second hand
- 2 chairs or desks of equal height

❸ Measure about 30 centimeters (12 in.) from the top of the cap and loop the string around the broomstick at that point. Tie the string tightly so that it doesn't slip off the stick. Then balance the broomstick between two chairs so that the bottle can swing freely by the string. The bottle should not hit the floor. If the chairs are not high enough, ask two friends to hold the ends of the stick.

❹ Get ready with your timer. You will count how many swings the pendulum makes in 10 seconds. Pull back the bottle and let it go. Count the swings and record the data in the chart below.

Length of String	Number of Swings (in 10 seconds)
30 centimeters	
15 centimeters	
45 centimeters	

❺ Untie the knot on the broomstick and measure 15 centimeters (6 in.) of string from the top of the cap. Retie the string on the broomstick. Predict: What will happen to the period of swing with the shorter string? Will the pendulum swing faster, slower, or the same number of times in 10 seconds? Repeat step 4, making sure to pull back the bottle from the same height. Record the data on the chart above.

❻ Untie the knot and get ready to repeat the experiment, this time with the string measuring 45 centimeters (18 in.) long. Based on your observations so far, what do you think will happen to the period of the swing with the longer pendulum? Why do you think so?

_____

❼ Repeat step 4 and record the results on the chart above. Did your results match your prediction? Based on your results, explain the relationship between the period of a swing and the length of the pendulum.

_____

Think About It: How can you find out if the mass of a pendulum affects the period of its swing? Devise an experiment and try it out.

Water Cycle in a Bottle

Get It Together

- clear 2-liter soda bottle (for each group of students)
- sharp scissors
- ruler
- wet sponge
- chalkboard or desktop
- glass of ice water
- "Water-Cycle Bottle" lab sheet (page 32)

Fizzy Science

All of the water on our planet has been recycled millions of times over. The *water cycle* (or *hydrologic cycle*) begins with *evaporation*. Energy from the sun (and wind) causes liquid water on the surface to change to gas (water vapor), which rises into the atmosphere. As water vapor rises, it cools until it *condenses* back to liquid water or freezes into solid ice crystals. When the ice or water droplets get heavy enough, they fall back to earth as *precipitation* and accumulate on the surface where the cycle starts again.

Before You Start

Cut the top off all the bottles, about 20 centimeters (8 in.) from the bottom (above the label). You'll have a cylinder and a funnel. Make sure the bottle caps are screwed tightly onto the funnel end of the bottle. Remove the labels.

What to Do

❶ Hold up a glass of water for the class to see and ask: Have you ever wondered where water comes from? *(Rain, faucet, a well, etc.)* Explain that the water we depend on for life is not new water, but has actually been used millions of times before. Even if it had been drawn from a deep well or collected from rain, the water has been recycled!

❷ Rub a wet sponge on a chalkboard to make a large wet spot. Invite several student volunteers to gently blow on the spot of water. Have the class observe the spot. Ask: What's happening to the water? *(It's disappearing.)* Explain that the water is actually going into the air. When water absorbs energy, it evaporates, turning from liquid to water vapor (a gas). Evaporation is the first step in the water cycle.

❸ Hold up the glass of ice water and invite students to touch it. Ask: How does the outside of the glass feel? *(Wet and cold)* Explain that the outside of the glass is wet because water vapor in the air that comes in contact with the cold glass loses energy and condenses back into liquid. Condensation leads to precipitation (rain or snow). Once the water returns to the Earth's surface, the water cycle starts again.

❹ Give each student a copy of the "Water-Cycle Bottle" lab sheet and invite students to make their own water-cycle bottle.

Water-Cycle Bottle

How does water move through the water cycle?

❶ Pour the cup of warm water into the soda bottle cylinder. Screw the top tightly onto the funnel top of the soda bottle and place it upside down into the soda bottle cylinder, as shown above. Use cellophane tape to seal the space where the two parts of the bottle join together.

❷ Place the bottle under direct sunlight for 5 minutes. Predict: What do you think will happen to the air temperature inside the bottle after it is exposed to the sun? Why?

_____

_____

❸ Observe the inner surface of the cylinder and the bottom of the funnel. Describe what you see.

_____

_____

❹ Place 2 or 3 ice cubes into the funnel and wait 5 minutes. Predict: What do you think will happen to the air temperature near the top of the soda bottle cylinder? Why?

_____

_____

❺ What's happening to the bottom of the funnel inside the soda bottle cylinder? Why?

_____

_____

Think About It: How does this activity model the way the water cycle works in nature?

You'll Need

- clear 2-liter soda bottle with top cut off about 20 cm from the bottom
- funnel top of the cut 2-liter bottle with screw cap
- 1 cup of warm water
- cellophane tape
- 2 or 3 ice cubes
- sunny windowsill
- watch or clock

Soda Bottle Science Scholastic Teaching Resources

Hot-Air Balloon

Get It Together

- clean, empty 2-liter soda bottle with cap (for each group of students)
- 2-liter soda bottle (for demo)
- round balloon (8–12 inches)
- electric hair dryer
- "The Invisible Crusher" lab sheet (page 34)

Fizzy Science

Even though we can't really see it, air is everywhere! Air is a collection of gases and, just like other forms of matter, gases change their volume when heated or cooled. Volume, along with mass and density, is a fundamental property of all matter. We calculate the density of a substance by using the formula Density = Mass / Volume (D=M/V). When air is heated, it expands (its volume increases) and its density decreases (the heated air becomes lighter). In the natural world, air is being heated and cooled all the time. When the sun shines on Earth, rocks and water absorb the light energy and heat up. In turn, these objects heat the air above them, causing its volume to increase and become lighter compared to the air around it. Since air is a fluid (it flows from one place to another), the heated air begins to rise, pushed up by the surrounding air, and new air flows in to take its place. This sideways-flowing air is what we call wind. If you trap the hot air in a bag or balloon, you can use the rising air to literally get a lift!

Before You Start

Remove the labels and plastic neck rings from all bottles.

What to Do

❶ Ask students: What causes the wind to blow? *(Changes in air temperature)* Explain that wind is really the result of the unequal heating and cooling of air. When the sun shines on Earth, it heats up the surface, which in turn heats the air above it. When air gets hot, it expands.

❷ Place the balloon over the neck of the bottle and call on a student volunteer. Have the student turn on the hair dryer and direct the flow toward the bottom of the bottle. Encourage the class to observe the balloon. Ask: What happens to the balloon as the air in the bottle gets hot? *(The balloon slowly starts to inflate.)*

❸ Explain that the balloon is inflating, not because the hair dryer is adding air to the bottle, but because the air in the bottle is getting hot. When air gets hot, it expands.

❹ Ask: What will happen to the balloon when you turn off the hair dryer? *(It will slowly deflate.)*

❺ Hand out copies of "The Invisible Crusher" lab sheet and invite students to see how else air changes when its temperature changes.

The Invisible Crusher

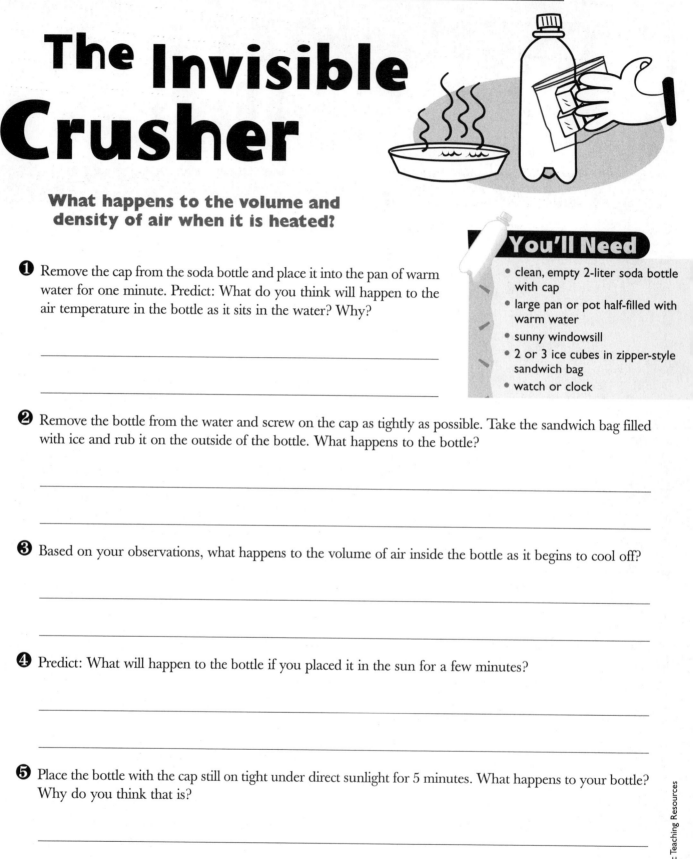

What happens to the volume and density of air when it is heated?

You'll Need

- clean, empty 2-liter soda bottle with cap
- large pan or pot half-filled with warm water
- sunny windowsill
- 2 or 3 ice cubes in zipper-style sandwich bag
- watch or clock

❶ Remove the cap from the soda bottle and place it into the pan of warm water for one minute. Predict: What do you think will happen to the air temperature in the bottle as it sits in the water? Why?

_____

_____

❷ Remove the bottle from the water and screw on the cap as tightly as possible. Take the sandwich bag filled with ice and rub it on the outside of the bottle. What happens to the bottle?

_____

_____

❸ Based on your observations, what happens to the volume of air inside the bottle as it begins to cool off?

_____

_____

❹ Predict: What will happen to the bottle if you placed it in the sun for a few minutes?

_____

_____

❺ Place the bottle with the cap still on tight under direct sunlight for 5 minutes. What happens to your bottle? Why do you think that is?

_____

_____

Think About It: What is the relationship between the volume of air and its temperature? What happens to the density of air as it gets warmer?

Soda Bottle Science Scholastic Teaching Resources

Whirlpool in a Bottle

Get It Together

- 2 clear 1-liter soda bottles (for each group of students)
- large, clear plastic cup half-filled with water
- sharp scissors
- spoon
- "The Twister" lab sheet (page 36)

Fizzy Science

Air and water are both *fluids*–their particles are free to move about and change their relative positions. Fluids always flow from a point of high pressure to a point of low pressure. All fluids also have a certain amount of friction. These frictional forces play a large role in how fluids move, and they often cause a fluid to form a *vortex*. A vortex is simply a circular pattern of flow around a central area of low pressure. If you've ever watched water draining down a sink, you've seen a vortex. In nature, vortices form all the time because of natural disruptions in fluid flow. A rock in a stream or a tree in the air will cause fluid to flow around it, triggering a vortex. Once the circular flow starts, friction in fluid amplifies the effect. The results can be spectacular! Tornadoes, hurricanes, and waterspouts are all examples of natural vortices.

Before You Start

Remove the labels and plastic neck rings from all bottles. Construct a model "twister" to show students (see page 36). Note: For best results, have students build their twister bottles one day and use them the next. This will give the glue on the bottles time to dry.

What to Do

❶ Call on a student volunteer and give him a spoon. Ask the student to start stirring the water in the clear plastic cup. Have the class observe what happens to the water.

❷ After stirring for about five seconds, have the student remove the spoon and walk around the class to give students a chance to observe the water in the cup. Ask: How is the water moving? *(It is flowing in a circular motion around the cup.)*

❸ Explain that the water flows in a circle due to a force called *friction*. Friction happens between things that rub together. When the spoon started moving through the water, it rubbed against the water and started the water moving.

❹ Ask: Why does the water keep moving even after the spoon is removed? *(Once the water starts moving, its molecules continue to rub against each other, so the flow continues.)*

❺ Explain that in the natural world, circular flow happens in fluids like water or air all the time. Ask: Can you think of an example where air naturally flows around in a circle? *(Tornadoes, hurricanes, dust devils)*

❻ Hand out copies of "The Twister" lab sheet and invite students to investigate how fluids flow in nature. Demonstrate how to construct the twister.

The Twister

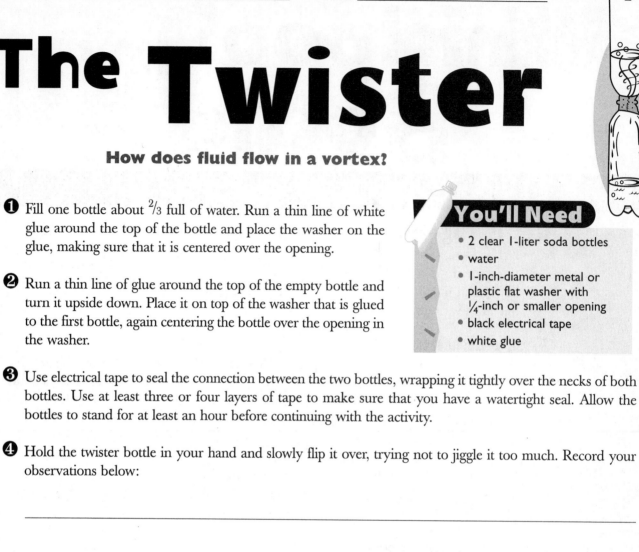

How does fluid flow in a vortex?

❶ Fill one bottle about ⅔ full of water. Run a thin line of white glue around the top of the bottle and place the washer on the glue, making sure that it is centered over the opening.

❷ Run a thin line of glue around the top of the empty bottle and turn it upside down. Place it on top of the washer that is glued to the first bottle, again centering the bottle over the opening in the washer.

You'll Need

• 2 clear 1-liter soda bottles
• water
• 1-inch-diameter metal or plastic flat washer with ¼-inch or smaller opening
• black electrical tape
• white glue

❸ Use electrical tape to seal the connection between the two bottles, wrapping it tightly over the necks of both bottles. Use at least three or four layers of tape to make sure that you have a watertight seal. Allow the bottles to stand for at least an hour before continuing with the activity.

❹ Hold the twister bottle in your hand and slowly flip it over, trying not to jiggle it too much. Record your observations below:

_____

_____

❺ Based on your observations, what's inside the lower bottle? How do you know?

_____

_____

❻ Predict: How can you get the water in the top bottle to flow faster into the bottom bottle?

_____

_____

❼ After all the water has drained into the lower bottle, flip the two bottles again, only this time, swirl the bottles so that the water in the top starts moving around in a circle. Record your observations:

_____

_____

Think About It: Why did the water flow faster when it was spinning? How does the movement in the twister bottle compare to the natural flow of air during storms?

Soda Bottle Science   Scholastic Teaching Resources

Rain Gauge

Get It Together

- clear 1-liter and 2-liter soda bottles (for each group of students)
- sharp scissors
- ruler
- measuring cup (1 pint or larger)
- large baking dish (9-by-12-in. or larger)
- "Make a Rain Gauge" lab sheet (page 38)

Fizzy Science

Collecting data using scientific instruments is important in *meteorology*, the study of weather. To record precipitation rates, meteorologists use a *rain gauge*. A basic rain gauge collects rainfall and measures how much rain has fallen over a period of time (measured in inches per day). Since raindrops falling from the sky are randomly spread out, rain gauges usually have a wide opening at the top and a funnel-like device that leads to a smaller tube inside. The smaller tube concentrates the rain in order to give a more accurate reading. Because the rain in the inner tube is concentrated, the measuring scale on the tube is expanded using a correction factor.

Before You Start

Remove the labels from all the bottles. For the 1-liter bottles, cut the top off about 20 centimeters (8 in.) from the bottom. For the 2-liter bottles, cut the top off about 15 centimeters (6 in.) from the bottom. Keep the 2-liter funnel-shaped tops. Construct a model rain gauge to show students (see page 38).

What to Do

❶ Have students recall information typically given in a weather report, such as how many inches of rain have fallen. Explain that precipitation is measured with a rain gauge.

❷ Fill a measuring cup with 500 milliliters (1 pint) of water. Pour the water in the baking dish. Say to students: *Let's say I put this pan outside in the rain and it rains only a little bit. If I put a ruler in the pan to measure the water depth, it will be hard to see how many inches of rain actually fell because the water in the pan will not be very deep.*

❸ Place the two soda-bottle cylinders in front of the class. Ask students: What will happen to the water's depth if I poured the water from the pan into the large cylinder? *(The water will rise to a higher level in the cylinder.)*

❹ Pour the water from the baking dish into the 2-liter bottle. Using the ruler, show the class how high the water level is now. Ask students: What would happen if I transferred the water from the wide cylinder into the 1-liter cylinder? *(The water level should go up even higher.)* Transfer the water to the 1-liter bottle.

❺ Explain that in rain gauges, the water is concentrated from a wide cylinder into a narrow cylinder to get a more accurate reading. For these devices to work correctly, however, the measuring scale on the inner cylinder must be calibrated to the wider cylinder. Hand out copies of "Make a Rain Gauge" to students.

Make a Rain Gauge

How does a rain gauge work?

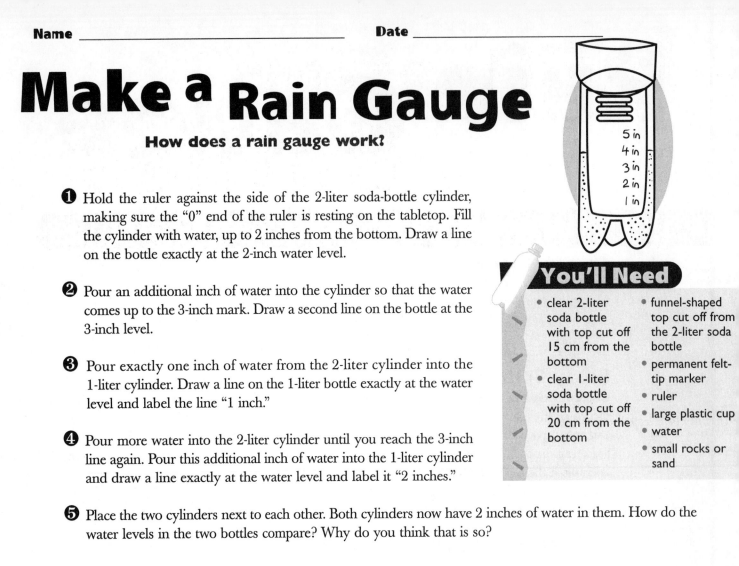

❶ Hold the ruler against the side of the 2-liter soda-bottle cylinder, making sure the "0" end of the ruler is resting on the tabletop. Fill the cylinder with water, up to 2 inches from the bottom. Draw a line on the bottle exactly at the 2-inch water level.

❷ Pour an additional inch of water into the cylinder so that the water comes up to the 3-inch mark. Draw a second line on the bottle at the 3-inch level.

❸ Pour exactly one inch of water from the 2-liter cylinder into the 1-liter cylinder. Draw a line on the 1-liter bottle exactly at the water level and label the line "1 inch."

❹ Pour more water into the 2-liter cylinder until you reach the 3-inch line again. Pour this additional inch of water into the 1-liter cylinder and draw a line exactly at the water level and label it "2 inches."

You'll Need

- clear 2-liter soda bottle with top cut off 15 cm from the bottom
- clear 1-liter soda bottle with top cut off 20 cm from the bottom
- funnel-shaped top cut off from the 2-liter soda bottle
- permanent felt-tip marker
- ruler
- large plastic cup
- water
- small rocks or sand

❺ Place the two cylinders next to each other. Both cylinders now have 2 inches of water in them. How do the water levels in the two bottles compare? Why do you think that is so?

_____

_____

❻ Repeat step 4 until you have transferred a total of 5 inches of water from the 2-liter cylinder into the 1-liter cylinder, making sure to mark and label every inch on the smaller bottle.

❼ The 1-liter cylinder now has a measuring scale calibrated to the wider opening of the 2-liter bottle. What is the advantage of having a wide opening to collect the rainwater but a narrow tube to measure it?

_____

_____

❽ Empty both cylinders then place the smaller cylinder inside the larger one. Invert the plastic funnel top from the 2-liter bottle onto the top of the bottle. Align the funnel so water will drain into the smaller cylinder. Your rain gauge is now ready! To use it, place it out in a clear open area before a rainstorm. To keep the rain gauge stable, you may want to fill the space between the two cylinders with small rocks or sand. When the rain has stopped, or after a set period of time, bring the rain gauge indoors. Take out the smaller cylinder and read how much rain has fallen.

Think About It: How might you redesign your rain gauge to get an even more accurate reading in the inner cylinder?

Soda Bottle Science Scholastic Teaching Resources

Infiltration Bottles

OBJECTIVE: **To investigate which earth material allows water to flow through it fastest**

Get It Together

- 3 clear 1-liter soda bottles (for each group of students)
- sharp scissors
- ruler
- large sponge
- brick or large flat rock
- several paper towels
- small cup of water
- "Percolation Test" lab sheet (page 40)

Fizzy Science

A major part of the water cycle involves *groundwater*–water that flows underground through *aquifers*, layers of porous rock sediment. Groundwater is a critical resource for millions of people who depend on wells for their water. When water is pumped out of a well, it reduces the amount of water in the aquifer. Fortunately, water flows from the surface back into an aquifer through a process called *infiltration*. The rate at which water flows through an aquifer depends on the rock's *permeability*. The size and interconnectedness of the rock's pore spaces affect its permeability. In general, the larger the pore spaces, the higher the rock's permeability and the faster water flows through it.

Before You Start

Remove the labels from all bottles. Cut the tops off the bottles about 18 centimeters (7 in.) from the bottom.

What to Do

❶ Guide students to imagine rain as it hits the ground. While some water forms puddles and runs off, much of it seeps into the ground. Explain that water that enters the ground helps recharge the groundwater system. Without it, wells would eventually run dry.

❷ Explain that before water can flow into an aquifer, it must first *infiltrate* or flow into the ground. The rate of infiltration (the rate at which water seeps into the ground) is affected by the permeability of the rock or soil.

❸ Call on a student volunteer. Place a sponge on a paper towel on a desk where everyone can see. Have the student slowly pour some water from a cup onto the sponge. Ask: What happens to the water when it hits the sponge? *(It seeps into the sponge.)*

❹ Next, place the brick or rock on a paper towel. Have the student slowly pour some water onto the brick. Ask: What happens to the water this time? *(Most of it runs off the surface.)* Ask: How does the permeability of the sponge compare to that of the brick? *(The sponge is much more permeable than the brick.)*

❺ Explain that while the brick and sponge represent extreme cases in permeability, different soils also have different permeabilities. Hand out copies of the "Percolation Test" lab sheet and invite students to test infiltration rates of different earth materials.

Percolation Test

Which soil type allows water to infiltrate the fastest?

❶ Place a soda-bottle cylinder on a table. Measure 10 centimeters (4 in.) from the bottom and draw a line on the bottle at this point. Repeat with the other two cylinders.

❷ Fill the first cylinder with gravel exactly up to the mark. As you put the gravel in, tap the soda bottle on the table several times to help the gravel settle. Repeat with the sand and the garden soil so that each material fills a different cylinder.

❸ Carefully observe the three materials in the bottles. How do they compare with each other?

_____

_____

You'll Need

- 3 clear 1-liter soda bottles with tops cut off 18 cm from the bottom
- 3-oz paper or plastic cup
- sink or other access to water
- ruler
- 1 liter of fine dry sand
- 1 liter of 1-inch-diameter gravel or small stones
- 1 liter of garden topsoil (not potting soil mix)
- watch or clock with second hand
- permanent felt-tip marker

❹ Based on your observations, which material do you think will have the highest permeability? Which will have the least? Why do you think so?

_____

_____

❺ Fill the small cup with water. Pour all the water into the cylinder filled with gravel and immediately start timing to see how many seconds it takes for all the water to seep into the gravel. Record the time below. Repeat the same procedure with the other cylinders, making sure you use the same amount of water each time.

Gravel: _____ seconds

Sand: _____ seconds

Soil: _____ seconds

❻ How did your results compare with your prediction?

_____

_____

Think About It: What soil properties would be most important for controlling the infiltration of water?

Soda Bottle Science Scholastic Teaching Resources

Porosity Bottle

Get It Together

- clear 1-liter soda bottle (for each group of students)
- clear 2-liter soda bottle (for each group of students)
- sharp scissors
- ruler
- 1 liter of sand
- 1 liter of garden topsoil (not potting soil mix)
- 1 liter of humus or organic potting soil
- 3 small plastic bowls (for each group of students)
- "Hold That Water!" lab sheet (page 42)

Fizzy Science

When it comes to storing water, not all soils are created equally! To maintain proper plant growth, soil must have a balanced level of moisture. Too much water, and plant roots get saturated and die; too little water, and they dry out. A soil's ability to hold moisture is controlled by its *porosity* and *permeability*. Porosity is a measure of how much total space there is between individual soil particles. In general, the more space between particles (the greater the porosity), the more water soil can store. Permeability depends on how big the pores are and how well they are connected. Soil with a great deal of sand and gravel holds very little moisture because their pores are large and water drains through them quickly. The best soils are ones that have large porosity and moderate permeability. These are soil mixtures with a wide range of grain sizes (sand, silt, and clay) plus a relatively high percentage of organic material (humus).

Before You Start

Cut the tops off all bottles right above their labels—about 15 centimeters (6 in.) from the bottom for 1-liter bottles and 20 centimeters (8 in.) from the bottom for 2-liter bottles. Keep only the funnel tops of the 2-liter bottles. Remove all the labels. Place about ½ cup of sand, garden soil, and humus into separate bowls for each student group.

What to Do

❶ Divide the class into groups of 4 to 6 students. Give each group a set of soil samples in bowls (sand, soil, and humus). Ask students to examine the three samples and record their observations about their color, texture, grain size, and composition.

❷ After 10 minutes, invite students to share their observations. Ask: Based on your observations, which material do you think will hold on to the most moisture? *(Humus)*

❸ Explain that the moisture-holding capacity of soil is controlled by two factors—*porosity*, which is the amount of space between soil particles, and *permeability*, which depends on how well the spaces or pores are connected. Hand out copies of the "Hold That Water!" lab sheet and invite students to conduct their own porosity tests on different types of soil.

Hold That Water!

Which type of soil is best at holding on to moisture?

❶ Carefully examine each of the three soil samples. Take a pinch of each dry sample in your fingers and rub it to feel the texture. Record your observations of each sample below:

Sand:

_____

Humus:

_____

Top Soil:

_____

You'll Need

- clear 1-liter soda bottle with top cut off 15 cm from the bottom
- 2-liter soda-bottle funnel top (without cap)
- pushpin or thumbtack
- ruler
- 1 liter of sand
- 1 liter of humus or organic potting soil
- 1 liter of garden topsoil
- 3 large (16-oz) plastic cups
- 6- to 8-oz plastic cup
- permanent felt-tip marker
- sink or other access to water
- watch or clock
- large plastic garbage bag
- paper towels

❷ Using the pushpin, punch 8 to 10 small holes in the bottom of the 1-liter bottle cylinder.

❸ Fill the cylinder with dry sand, up to 10 centimeters (4 in.) from the bottom. Use the marker to label a large plastic cup "Sand." Set the funnel top on the cup then place the 1-liter cylinder into the wide end of the funnel, as shown above. Fill the small plastic cup to the top with water.

❹ Looking at the clock or watch, prepare to pour the water into the top of the cylinder. Slowly pour the water into the sand, making sure to spread the water evenly around the top of the sample. As the water percolates through the sand, it will flow out the bottom of the cylinder and into the funnel, collecting in the plastic cup. Allow the water to flow for exactly 2 minutes and then remove the funnel from the cup.

❺ Empty the wet sand from the cylinder into the large plastic garbage bag. Rinse out the cylinder to get rid of all the sand particles. Make sure there are no particles blocking the small drain holes at the bottom of the cylinder. Dry the inside of the cylinder and the funnel with a paper towel.

❻ Repeat steps 3 to 5 using the humus, labeling the large plastic cup "Humus." Then repeat the same steps using the garden topsoil.

❼ After you have tested all three samples, compare the water level in the three cups. Which sample allowed the most water to flow through? Which sample held on to the most water?

_____

_____

Think About It: What type of soil holds on to moisture best? Why do you think that is so?

Sediment-Settling Bottles

Get It Together

- 2 clear 1-liter soda bottles (for each group of students)
- sharp scissors
- ruler
- 1 cup builders or beach sand
- 1 cup garden topsoil (not potting soil)
- 2 small plastic bowls (for each group of students)
- "Sediment Sorting" lab sheet (page 44)

Fizzy Science

Most soil is composed of broken bits of rocks and minerals called *sediment*. Sediment comes in four size groups—gravel, sand, silt, and clay. While individual pieces of gravel are large enough to be examined with the naked eye, clay-sized particles are so small they cannot be seen even with the best light microscopes. The proportion of each type of sediment present in soil determines the soil's porosity, permeability, infiltration rate of water, and moisture-storage capacity. A simple way of sorting sediment is by pouring soil through a column of water and allowing the water to separate out the different components. In general, the biggest particles sink first and the smallest particles stay in suspension the longest. Organic material floats. By comparing the relative thicknesses of the different sediment layers, soil scientists can approximate how a soil is going to behave.

Before You Start

Cut the tops off all the soda bottles right above the labels—about 18 centimeters (7 in.) from the bottom. Remove all labels. For each group of students, prepare a bowl filled with one cup of garden soil and another bowl filled with one cup of sand. You might want to have students bring in a variety of soil samples to compare with one another.

What to Do

❶ Divide the class into groups of 4 to 6 students. Give each group a bowl of soil and a bowl of sand. Ask students to examine the two samples and record their physical properties, such as color, texture, grain size, and composition.

❷ After 10 minutes, invite students to share their observations. Ask: How are the two samples the same? How are they different? (*Both samples contain minerals. The soil is darker than the sand, and the sand is more uniform in texture.*)

❸ Explain that most soils are made up of mineral grains called *sediment*, which comes in many sizes. The three main types of sediment usually found in soil are sand, silt, and clay. Different soils have different mixtures of these sediments and, from an agricultural standpoint, the best soils have a mix of all three.

❹ Hand out copies of the "Sediment Sorting" lab sheet and invite students to see how different soils "stack up."

Sediment Sorting

How can water be used to sort out sediment?

❶ Fill both soda-bottle cylinders with water, up to 12 centimeters (5 in.) from the bottom. Fill one small cup with sand and the other with garden soil. Closely examine both the sand and soil samples. Take a pinch of each in your fingers and compare their texture. Record your observations below:

Sand: _____

_____

Soil: _____

_____

You'll Need

- 2 clear 1-liter soda bottles with tops cut off 18 cm from the bottom
- 1 cup builders or beach sand
- 1 cup garden topsoil (not potting soil)
- 2 3-oz plastic or paper cups
- sink or other access to water
- watch or clock
- ruler

❷ Take the cup with the sand and quickly pour it into one of the cylinders. Observe what happens when the sand hits the water. Continue to observe the sand for about 10 minutes. What happens to the sand? What happens to the water?

_____

_____

❸ Take the cup of soil and prepare to pour it into the other cylinder. But first, predict how the soil will behave when it hits the water. Write your predictions here:

_____

_____

❹ Quickly pour the soil sample into the second cylinder. Observe what happens for about 10 minutes. How does this sample compare to the sand sample in the first cylinder? How does the water compare with the water in the cylinder with the sand sample? Which sample shows a greater separation of layers? Why do you think that is so?

_____

_____

Think About It: How might you use this technique to compare the composition of different types of soil?

Soda Bottle Science   Scholastic Teaching Resources

Water-Pressure Bottle

Get It Together

- 1 or 2-liter soda bottle (for each group of students)
- sharp scissors
- balance scale or lab scale
- large empty plastic cup
- "Pressure Streaming" lab sheet (page 46)

Fizzy Science

Anyone who has ever dived deep into a pool or pond has felt the effects of water pressure. The deeper you go, the more you can feel water pressing against your body, especially your ears. Water has pressure because, like all forms of matter, it has mass. As you dive underwater, the volume of water above you increases as well as its mass. At some point, the pressure would become so great that it could crush you! To reach the maximum depths of the ocean, scientists use special submarines that can withstand enormous pressures under the sea. Incredibly, even at depths that would crush a regular submarine, biologists have found living creatures that are alive and thriving. In many cases, these organisms have special adaptations that help them survive the pressure.

Before You Start

Remove the labels from all soda bottles.

What to Do

❶ Ask students to think back to the last time they went swimming. Ask: When you dived underwater, did you notice how your body, particularly your ears, felt? (Guide students to verbalize the sensation of water pushing down on them.)

❷ Explain that the reason water felt like it was pressing down on them was because, like all matter, water has mass.

❸ Ask a student volunteer to assist you with a demonstration. Place a large empty cup on top of a scale or balance and ask the student to begin pouring water into the cup. Have the rest of the class observe the scale. Ask: What happens to the weight of the cup as we add more water to it? *(It increases.)*

❹ Explain that as you dive underwater, there is a greater volume of water above you so that the weight or pressure of the water pressing on your body increases. By using a simple test apparatus, students can determine just how much water pressure increases with depth.

❺ Hand out copies of the "Pressure Streaming" lab sheet and invite students to do their own water-pressure tests.

Pressure Streaming

What is the relationship between water pressure and depth?

❶ Place the ruler next to the soda bottle. Starting at the bottom, draw a line every inch on the bottle until you reach the 7-inch mark. Use the pushpin or thumbtack to punch a small hole in the side of the bottle one inch from the bottom.

❷ Place a small piece of tape over the hole and put the bottle inside the large pan. Fill the bottle with water up to the 2-inch mark, exactly one inch above the hole.

❸ Predict: What will happen when you pull the tape off the bottle and open the hole?

_____

❹ Remove the tape from the hole and observe what happens. After the water level in the bottle drops below the hole, dry the outside of the bottle with the paper towel and replace the tape over the hole. Refill the bottle with water, this time up to the 3-inch mark (2 inches above the hole). Predict: What will happen to the water when you pull the tape off the bottle this time?

_____

❺ Repeat step 4 four more times, each time filling the bottle with another extra inch of water. Use the ruler to measure the distance the water flows out of the hole at the moment you remove the tape. Measure from the bottle to the spot where the water lands in the pan. Record the data below:

Water Level in Bottle	Water Distance from Bottle
4-inch mark	_____ inches
5-inch mark	_____ inches
6-inch mark	_____ inches
7-inch mark	_____ inches

❻ Explain the relationship between the water level in the bottle and the distance the water flows out of the hole and into the pan.

_____

_____

Think About It: What is the relationship between water depth and water pressure?

Soda Bottle Science   Scholastic Teaching Resources

You'll Need

- 1 or 2-liter soda bottle
- 9-by-12-inch aluminum pan or plastic dish pan
- ruler
- permanent felt-tip marker
- small piece of electrical or duct tape
- pushpin or thumbtack
- paper towel
- access to water

Soda Bottle Terrarium

Get It Together
- 2 clear 2-liter bottles (for each group of students)
- sharp scissors
- ruler
- "Life in a Bottle" lab sheet (pages 48–49)

Fizzy Science

An *ecosystem* is a naturally functioning system that includes a community of living things and all the other environmental components that allow those organisms to carry out their life processes. Included in the ecosystem are many *abiotic* (not produced by living organisms) factors—such as water, oxygen, carbon dioxide, soil, and nutrients. One way to understand how an ecosystem functions is to make a model of one, like a *terrarium*. The word *terrarium* comes from the Latin word *terra*, which means "earth." A typical terrarium contains soil, plants, and a host of other living things, ranging from insects and snails to worms and microbes. Terrariums can be either open or closed systems. In an open terrarium, air is free to circulate in and out of the system and moisture is added regularly to sustain the various life processes. In a closed system, moisture is added at the beginning when the terrarium is first set up but then the system is sealed and allowed to run on its own.

Before You Start

Remove the labels from all bottles. Cut off the tops so that half of the bottles are 23-centimeter (9-in.) cylinders and half are 15-centimeter (6-in.) cylinders.

What to Do

❶ Tell students that they are going to create their own little world! They will be building a mini-ecosystem, complete with all of the natural factors needed to keep it going.

❷ Ask: What does the word *ecosystem* mean? *(An ecosystem is a part of the environment that includes living things and all the factors that they need to survive.)*

❸ Challenge the class to brainstorm all the things that make up a typical forest ecosystem, such as soil, plants, trees, insects, air, and water. List their responses on the board.

❹ Explain that even though some of these factors may seem less important than others, they all play a role in keeping an ecosystem going. Sometimes just a change in one small factor can have an enormous effect on an entire ecosystem. Since it's difficult to study large-scale ecosystems in nature, scientists often create model ecosystems that allow them to test different factors.

❺ Explain that one way to model an ecosystem is to build a terrarium, which is like an aquarium, but has soil and plants instead of water. Hand out copies of the "Life in a Bottle" lab sheet and invite students to construct their own ecosystem in a bottle.

Life in a Bottle

How does a terrarium model an ecosystem?

❶ In this activity you are going to build a simple terrarium, which is really a mini ecosystem in a bottle. During your investigation, you will be able to control different factors to see how they impact different organisms living in your ecosystem.

❷ Take the shorter of the two soda-bottle cylinders and fill it with about $7\frac{1}{2}$ centimeters (3 in.) of moist garden topsoil. Sprinkle $\frac{1}{4}$ cup of wild birdseed or grass seed on the soil and then add another $2\frac{1}{2}$ centimeters (1 in.) of soil on top of the seed. Spray the top of the soil with water until it just starts to form little puddles.

You'll Need

- clear 2-liter bottle with top cut off 15 cm from the bottom
- clear 2-liter bottle with top cut off 23 cm from the bottom
- $\frac{1}{2}$ liter garden topsoil
- $\frac{1}{4}$ cup wild birdseed or grass seed
- water
- plant mister or spray bottle
- sunny windowsill

❸ Take the taller soda bottle cylinder and turn it upside down over the one with the soil in it. The large cylinder should be a little narrower than the shorter one. Slip the open end of the tall cylinder into the short cylinder and push it down until it fits tight.

❹ Observe the terrarium from the side. What do you see happening to the water that you sprayed on top of the soil? What part of the environment does the water represent?

_____

_____

❺ Place the terrarium on a sunny windowsill. Check it each day for seven days and note any changes that you see in the system on the chart below:

Day	Observations
1	
2	

Soda Bottle Science Scholastic Teaching Resources

Day	Observations
3	
4	
5	
6	
7	

❻ Once your ecosystem has become established and the plants are growing, you can begin testing different conditions. Here are some of the variables that you can test:

- **Population size:** What would happen if you added more seeds to the soil?
- **Population variation:** What would happen if you added animals, such as insects and worms, to the terrarium?
- **Temperature:** What would happen if you placed the terrarium in a cool spot?
- **Sunlight:** What would happen if you placed it in the dark?
- **Water:** What would happen if you added more water or took the top off so it will dry out?
- **Acid rain:** What would happen if you added a teaspoon of vinegar or lemon juice to the water?

Think About It: How do terrariums help scientists determine what factors impact the health of an ecosystem?

Soda Bottle Bug Habitat

OBJECTIVE: To observe the behavior and life cycle of isopods

Get It Together
- clear 2-liter soda bottle with cap (for each group of students)
- several pill bugs (purchased or collected in the wild)
- magnifiers
- paper plates
- "Pill Bug Palace" lab sheet (page 51)

Fizzy Science
Isopods are fascinating creatures that belong to the class *Crustacea*. One of the most common isopods is the pill bug, which is not really a bug at all. Although often confused with insects, pill bugs (as all isopods) are more closely related to crabs. Pill bugs can be found in moist soils, usually under rocks or decaying logs. They are about 1¼ centimeters (½ inch) long and have a flattened body. They breathe through gills and have seven pairs of legs. Their most striking feature is the armor-like plates that cover their bodies. When threatened, pill bugs roll up into a ball, making it hard for birds and other animals to eat them. The life span of a typical pill bug is about two years but they can live up to five years. They are decomposers, feeding mostly on decaying plant and animal matter. In captivity, pill bugs feed on fruit and potatoes.

Before You Start
Remove the labels from all soda bottles. If possible, organize a class field trip to collect pill bugs. They are commonly found under rocks and rotting wood and around building foundations where it is dark and damp. Or you can order them from a science supply house. Store the pill bugs in a soda bottle with some moist soil. Punch small holes in the bottle to let air in.

What to Do
❶ Divide the class into small groups and give each group a set of magnifiers and a paper plate. Call on a representative from each group to come forward with the paper plate. Give each group some soil with several pill bugs. Explain that the creatures are called *isopods*, which means "similar feet." Their common name is pill bug.

❷ Have students use the magnifiers to carefully observe the creatures and then make a list of their physical characteristics, including the number of legs, number of body segments, presence of wings or antennae, and so on.

❸ With the class's input, list the pill bug's physical characteristics on the board. Ask: What group of animals do you think pill bugs belong to? (*Crustaceans*)

❹ Explain that even though pill bugs look like insects—and they have the word "bug" in their name—isopods really belong to a class of animals called crustaceans. Ask students to name some other crustaceans. (*Crabs, lobsters, shrimp*)

❺ Hand out copies of the "Pill Bug Palace" lab sheet, inviting students to construct their own pill-bug habitat.

Pill Bug Palace

**What are isopods?
How do they behave?**

You'll Need

- clear 2-liter soda bottle with cap
- sharp scissors
- 15-by-10-cm sheet clear plastic wrap
- 2 large rubber bands
- 1 liter forest soil (or humus or topsoil from garden center)
- dead leaves
- 3 or 4 pill bugs
- sunny windowsill or desk lamp
- apple, raw potato, and other types of food
- spray bottle with water

❶ Turn the 2-liter bottle on its side. Using sharp scissors, carefully cut a small rectangular opening (about 10 by 5 centimeters) on one side of the bottle.

❷ With the bottle still lying on its side, fill the bottom of the bottle with moist organic soil to a depth of about 5 centimeters (2 in.), making sure to keep the opening facing up. Spread the soil evenly along the bottom. Place three or four pill bugs into the soil and add a few dead leaves.

❸ Stretch a piece of clear plastic wrap over the opening and hold it in place with two rubber bands. Your habitat is complete and ready for observation.

❹ Observe the pill bugs in their new home. Where do they go? Why?

_____

_____

❺ Place the habitat on a sunny windowsill or under a bright light. How do the pill bugs react?

_____

_____

❻ Cut up a piece of apple or raw potato and place it in the habitat. How do the pill bugs react? How do they eat? Try other types of food like carrots, bananas, and so on. What is their favorite food?

_____

_____

❼ To stay healthy, pill bugs need to be fed and kept moist every few days. Use a spray bottle to thoroughly soak the top of the soil.

Think About It: What characteristics do pill bugs share with insects and worms? How are they similar to other crustaceans?

Soda Bottle Greenhouse

Get It Together

- clear 2-liter soda bottle (for each group of students)
- green 2-liter soda bottle (for each group of students)
- sharp scissors
- ruler
- lima beans or grass seeds
- soil
- small plastic cups
- flashlight
- "Growing Green" lab sheet (pages 53–54)

Fizzy Science

To grow, most plants need water, air, nutrients from soil or water, and light. Plants get their energy directly from light. During a process called *photosynthesis*, they use light energy to manufacture simple sugars from water and carbon dioxide. Without light, most plants would die. But not all light is created equally. Sunlight, which most plants rely on, is a full-spectrum light source. It consists of a wide range of different wavelengths of light energy. You can see these different wavelengths when you look at a rainbow. Other light sources, such as incandescent or fluorescent bulbs, frequently are missing some wavelengths of light. As a result, indoor plants that get their energy from an artificial lamp or from sunlight passing through a tinted window (which filters out certain wavelengths) are often stunted.

Before You Start

Cut the bottoms off all the bottles, about 25 centimeters (10 in.) from the top. Remove all labels. Before beginning the activity, you may want to grow lima beans or grass seeds in small plastic cups for students to use. Or, you can purchase seedlings from a local garden center.

What to Do

❶ Ask students: What do plants need to grow? *(Water, soil, air, light)* Explain that unlike animals that get their energy from eating other things, plants use sunlight to make their own food through a process called *photosynthesis*.

❷ Ask: Have you ever noticed that indoor plants sometimes don't do as well as the same plants grown outside? While many factors could cause plants to grow poorly, one problem is that sometimes light passing through a window is filtered by the glass.

❸ Darken the room. Turn on a flashlight, directing it toward a wall to make a spot of light. Next, shine the light through the clear plastic bottle. There should be little change in the spot of light.

❹ Hold up the green bottle and ask: What will happen to the light as it passes through the green bottle? *(The light will have a greenish tint.)*

❺ Explain that the green plastic blocks all the colors of light except for green. Ask: Do you think a plant will grow the same way under green light as under white light? Hand out copies of the "Growing Green" lab sheet and invite students to find out!

Name _____   Date _____

Growing Green

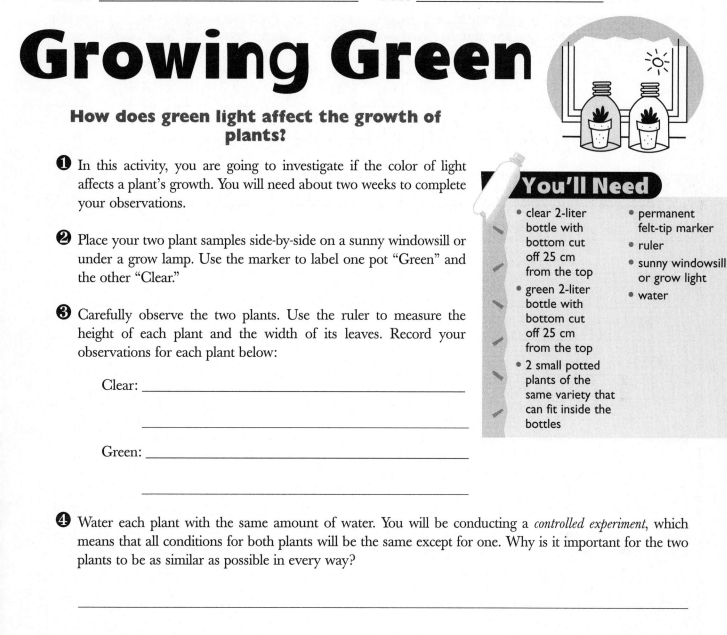

How does green light affect the growth of plants?

❶ In this activity, you are going to investigate if the color of light affects a plant's growth. You will need about two weeks to complete your observations.

❷ Place your two plant samples side-by-side on a sunny windowsill or under a grow lamp. Use the marker to label one pot "Green" and the other "Clear."

❸ Carefully observe the two plants. Use the ruler to measure the height of each plant and the width of its leaves. Record your observations for each plant below:

Clear: _____

_____

Green: _____

_____

You'll Need

- clear 2-liter bottle with bottom cut off 25 cm from the top
- green 2-liter bottle with bottom cut off 25 cm from the top
- 2 small potted plants of the same variety that can fit inside the bottles
- permanent felt-tip marker
- ruler
- sunny windowsill or grow light
- water

❹ Water each plant with the same amount of water. You will be conducting a *controlled experiment*, which means that all conditions for both plants will be the same except for one. Why is it important for the two plants to be as similar as possible in every way?

_____

_____

❺ Cover the plant labeled "Green" with the green soda bottle cylinder and the plant labeled "Clear" with the clear soda bottle cylinder. Leave the caps off the bottles. Check the plants each day over a 14-day period, remembering to water them every few days so they don't dry out. Record your observations below:

Day	Green	Clear
1		
2		
3		

(continued)

Day	Green	Clear
4		
5		
6		
7		
8		
9		
10		
11		
12		
13		
14		

❻ Based on your observations, did the different-colored lights have any impact on the plants' growth? In what way?

_____

_____

Think About It: Do you think all plants would react the same way in this experiment? How might you check your hypothesis?

Transpiration Bottle

OBJECTIVE: To observe how transpiration works in plants

Get It Together

- 2 clear 2-liter bottles (for each group of students)
- sharp scissors
- ruler
- lima beans or grass seeds
- soil
- small plastic cups
- fresh green tree leaves
- magnifiers
- "You're All Wet!" lab sheet (page 56)

Fizzy Science

Most plants need to be watered regularly to stay healthy. Water is used in photosynthesis to make food for the plant. It also transports minerals, nutrients, and food throughout the plant. Most of the water that a plant takes in at its roots is lost back out through its leaves in a process called *transpiration*. Transpiration helps keep a plant cool and helps concentrate essential elements inside the plant tissue. The underside of a leaf has tiny openings called *stomata*, which allow gases to pass in and out of the plant. This is where most transpiration takes place. The topside of a leaf has no openings and is often covered with a thick waxy layer called the *cuticle*. This layer prevents too much transpiration from taking place. Many times when the sun is shining brightly and the stomata are open, water loss can be significant. During evening hours, the stomata close so plants can recharge.

Before You Start

Cut off the bottle tops so that half of the bottles become 10-centimeter (4-in.) tall cylinders and half become 23-centimeter (9-in.) tall cylinders. Before the activity, you may want to grow some lima beans or grass seeds in small plastic cups for students to use. Or, you can purchase seedlings from a local garden center.

What to Do

❶ Ask students: Why do plants need water to grow? *(Water is important for making food and for transporting nutrients in the plant.)* Explain that water also helps keep plants cool. On hot days, water evaporates off the leaves in a process called *transpiration*.

❷ Pass around the leaf samples and magnifiers. Have students examine both sides of the leaf closely. Ask: How does the topside of the leaf compare to the bottom? *(The topside is smooth, darker, and feels a little waxy; the bottom feels rough and has little holes.)*

❸ Explain that leaves on most plants are designed to minimize water loss. Point out the tiny openings at the bottom of the leaves. Explain that they are called *stomata* and are like the pores in skin, where much water comes out.

❹ Ask: Why would it be bad for the stomata to be on the topside of the leaves? *(If the openings were in direct sunlight, the plant would lose too much water and die.)* Hand out copies of the "You're All Wet!" lab sheet and invite students to investigate transpiration.

You're All Wet!

What factors affect transpiration in plants?

❶ Make sure your plant is watered well. Place the plant inside the shorter soda bottle cylinder. Take the taller cylinder and turn it upside down over the first. Slip the tapered end of the taller cylinder inside the shorter one and push them together so they fit snugly together. Why do you think the two cylinders must fit together tightly?

_____

_____

You'll Need

- clear 2-liter soda bottle with top cut off 10 cm from the bottom
- clear 2-liter soda bottle with top cut off 23 cm from the bottom
- small potted plant
- paper towel
- large cardboard box or big brown paper bag
- sunny windowsill or grow light
- watch or clock

❷ Place the bottle with the plant on a sunny windowsill or under a grow light. Let it stand in the light for about 30 minutes. Observe the inside of the bottle and record what you see. Why do you think this is happening?

_____

_____

❸ Next, take the two cylinders apart and dry the inside of the bottles with a paper towel. Replace the plant in the bottles as you did in step 1. Cover the plant with a large cardboard box or brown paper bag and move it from the windowsill to simulate nighttime. Predict: What will happen to the moisture level in the bottle this time? Why do you think so?

_____

❹ After 30 minutes, remove the covering from the plant and observe the inside of the bottle. How does the inside of the bottle compare with the way the bottle looked in step 2? Why do you think this is so?

_____

_____

❺ Based on your experiments, when would you say that plants transpire the most? How does this help or hurt them?

_____

_____

Think About It: Do you think that all plants transpire at the same rate? How might you design an experiment to check your hypothesis?

Soda Bottle Science Scholastic Teaching Resources

Colored Leaves

Get It Together

- clear 1-liter soda bottle (for each group of students)
- sharp scissors
- ruler
- cross-sectional piece of a tree trunk showing the rings (or photo)
- carrots
- magnifiers
- plastic knives
- "Leaf of a Different Color" lab sheet (page 58)

Fizzy Science

Most plants would be in big trouble if not for their stems! Stems transport water, food, and nutrients throughout the plants. Inside the stem are two distinct passageways that allow these critical substances to circulate around the plant. The first channel is called the *xylem*, which carries water and dissolved minerals from the roots up to the leaves. In woody plants, like trees, the xylem tissue is quite strong and provides most of the support for the plant. In perennials, a new layer of xylem grows each year, so if you cut across the stem, you can see each successive layer as a set of growth rings. The second transport channel is called *phloem*, which carries food in the form of sugars from the leaves to the rest of the plant.

Before You Start

Cut the tops off all the bottles, about 15 centimeter (6 in.) from the bottom. Contact your local parks department to try and get a cross-section of a tree trunk. Or find a photo of one.

What to Do

❶ Tell students: *In humans and animals, blood flows through veins and arteries to each organ. Do plants have blood vessels too?* Explain that instead of blood vessels, plants have special cells that act like pipelines, allowing water and nutrients to flow around the plant.

❷ Divide the class into small groups and pass around the carrots, plastic knives, and magnifiers. Ask: What part of the plant is the carrot? *(Root)* Have students cut across the carrot and observe the pattern inside. Ask: What shape do you see inside the carrot? *(An inner circle of tissue)* Explain that this inner circle is called the *xylem*, and it's the passageway through which water gets from the roots to the rest of the plant.

❸ Pass around the tree cross section or a photo of one. Ask: What part of a plant is a tree trunk? *(Stem)* What shapes do you see in the trees? *(Rings)*

❹ Explain that tree rings are also xylem. Each year, woody plants grow a new layer of xylem right under the bark and the old layer dies. By counting the layers of xylem, you can tell a tree's age. Non-woody plants also have xylem running through their stems.

❺ Hand out copies of the "Leaf of a Different Color" lab sheet and invite students to find a plant stem's xylem.

Leaf of a Different Color

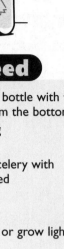

How do plants transport water inside them?

❶ Take a stalk of celery and examine it carefully on the outside. With a plastic knife, cut across the stalk near the bottom and examine the inside of the stem, too. What part of the stem do you think transports the water? Record your observations below:

_____

_____

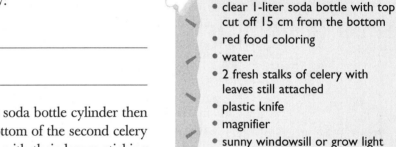

You'll Need

- clear 1-liter soda bottle with top cut off 15 cm from the bottom
- red food coloring
- water
- 2 fresh stalks of celery with leaves still attached
- plastic knife
- magnifier
- sunny windowsill or grow light
- watch or clock

❷ Put a few drops of red food coloring in the soda bottle cylinder then fill it halfway with water. Cut across the bottom of the second celery stalk and place both stalks in the cylinder, with their leaves sticking out the top.

❸ Place the cylinder on a sunny windowsill or under a grow light and wait about 15 minutes. Predict: What changes will you see in the plants?

_____

_____

❹ After 15 minutes, remove one of the celery stalks and, with the knife, cut across the stem about $2\frac{1}{2}$ centimeters (1 in.) from the bottom. Look at the bottom of the stem and describe what you see:

_____

_____

❺ Based on your observations, what part of the stem is the xylem? How do you know?

_____

_____

❻ Leave the second stalk of celery in water overnight. Based on your observation of the stem, what do you think will happen to the leaves? Write your prediction. Then record your observations the next day:

_____

_____

Think About It: How might you use this technique to identify the xylem tissue in other plants?

Soda Bottle Science   Scholastic Teaching Resources

Yeast Beast

OBJECTIVE: To investigate how yeast cells get food

Get It Together

- 2 clear 1-liter soda bottles (for each group of students)
- sharp scissors
- matzo
- regular white bread
- packets of yeast
- magnifiers
- paper towels
- "A Fungus Among Us" lab sheet (page 60)

Fizzy Science

We owe a great deal to yeast! Without yeast, all bread would be flat and hard like matzo. In its dormant state, yeast looks like brown dust, yet it is very much alive! Yeast is a microscopic, one-celled organism that reproduces by means of spores and buds. It is a type of fungus and usually gets its nourishment from dead or decaying organic matter. Yeast feeds by causing a chemical reaction in sugars and, in the process, breaks down the sugar to produce alcohol. It is often used in brewing beer and making wine. As yeast "eats," it releases carbon-dioxide gas, creating bubbles. These bubbles cause bread dough to rise when yeast is used for baking. Yeast thrives in warm and damp environments. When it's cold and dry, yeast goes dormant as it waits for the right conditions.

Before You Start

Remove the labels and cut the plastic neck rings off the bottles.

What to Do

❶ Poll students by asking: How many of you like the smell of bread baking? It's one of the best smells on earth, and we owe it all to a tiny organism called yeast! Explain that besides making bread smell good, yeast also makes it rise.

❷ Divide the class into small groups and give each group a piece of matzo and a piece of white bread. Have students examine each piece. Ask: What are some properties of each piece of bread? *(The matzo is hard and flat and has no special smell; the bread is soft and fluffy, has lots of holes, and smells sweetish.)* Explain that the difference between the two types of bread can be attributed to yeast. The white bread was made with yeast, while the matzo was not.

❸ Give each group a packet of yeast, a paper towel, and some magnifiers. Have students open the packet and sprinkle some yeast on the paper towel. Ask them to observe the yeast with the magnifier and describe some of its properties.

❹ Ask: Is yeast a living thing? *(Yes)* Explain that the yeast is not dead, but rather, dormant–it is in a suspended state, waiting for the right environmental conditions. Hand out copies of the "A Fungus Among Us" lab sheet and invite students to experiment with yeast.

A Fungus Among Us

What do yeast cells need to live?

❶ In this activity, you are going to find out what environmental conditions are ideal for yeast. Open the packet of yeast and look at it closely. What does the yeast look like? What properties does it have? Record your observations below:

_____

_____

You'll Need

- 2 clear 1-liter soda bottles
- 2 packets of dry active yeast
- 1 sugar packet
- warm water
- 2 round balloons (10–12 inches)
- permanent felt-tip marker

❷ When yeast first comes out of the packet, it is dormant. To activate it, you must first add water. Fill each bottle about halfway with warm (not hot!) water. Add one packet of yeast to each bottle and gently swish each bottle around to mix. Add a packet of sugar to one of the bottles and mark the bottle with an "S." Swish the bottle around a little to mix the sugar into the water.

❸ Place a balloon over the mouth of each bottle. Place the bottles in a safe location that is relatively warm. You are going to observe the two bottles over four days. Write your observations on the chart below:

Day	Bottle with Sugar	Bottle without Sugar
1		
2		
3		
4		

❹ Based on your observations, which bottle showed more activity? Why do you think this was so? Write your ideas below:

_____

_____

❺ Based on your observations, why do you think bread baked with yeast gets soft and fluffy while bread baked without yeast is hard and flat?

_____

_____

❻ How might you test other environmental conditions to see how they affect the activity of yeast? How might you test hot and cold? Light and dark? Different sources of food?

60

Think About It: Why do you think many bread recipes require sugar, corn syrup, or molasses?

Soda Bottle Science Scholastic Teaching Resources

Soda Bottle Composters

Get It Together

- 4 clear 2-liter soda bottles (for each group of students)
- sharp scissors
- ruler
- magnifiers
- small paper plates
- plastic spoons
- 2 to 3 liters of natural topsoil
- "Let It Rot" lab sheet (pages 63–64)

Fizzy Science

If you've ever taken a close-up look at natural forest soil, you know we live in a rotten world! While most soils are composed of minerals weathered from rocks, it's the organic matter derived from rotting plants and animals that really make soil productive. In a natural system, nutrients continuously get recycled through a process called *decomposition*. As organic material, such as leaf litter, collect on the forest floor, a whole host of organisms breaks it down into simpler compounds. Decomposition is a critical part of all natural ecosystems. Without it, the next generation of *producers* (plants) and *consumers* (animals) would lack many of the essential elements they need to carry out their life cycles. The organisms that are responsible for decomposition are called *decomposers*, and they come in many shapes and sizes. They include microscopic organisms, like bacteria and fungi, as well as insects, worms, snails, and even larger animals. Like all living things, *decomposers* have certain needs, like water and air. Temperature also plays a big role in decomposition, which happens mostly in spring and summer, when the soil is fairly warm. While some decomposers can work at temperatures below freezing, decomposition generally comes to a standstill during the winter months.

Before You Start

Remove the labels from all bottles. Each group of students will need two sets of compost chambers, with each chamber made up of two bottles. Use scissors to cut the tops off so that half of the bottles are 10 centimeters (4 in.) tall and half are 23 centimeters (9 in.) tall. The taller cylinder should be cut off at the point just above where the top of the bottle begins to narrow. If you have a wooded area near your school, you may want to do the first part of the activity in the field. That way, students can see the actual environment associated with natural decomposition in the soil. If this is not possible, gather soil samples, leaf litter, and grass clippings in advance.

What to Do

❶ Ask students if they have ever taken a close-up look at natural soil. Ask: What kind of stuff is in natural soil? *(Rocks, leaves, twigs, bugs, and so on)* Explain that even though many people take dirt for granted, natural soil is really a very complex environment.

❷ Divide the class into small groups and give each group a paper plate, a plastic spoon, and magnifiers. Place a large scoop of soil on each plate and ask students to list what they see in the soil.

Soda Bottle Composters

(continued)

❸ After a few minutes, invite students to call out what they've found as you list the various soil components on the board. Point out that if they look at the list closely, most of the components can be broken down into two main groups—*organic materials* (living and dead things, like leaves, roots, and insects) and *inorganic materials* (small rocks and grains of mineral).

❹ Ask: How does the organic part of soil change over time? *(They eventually decompose or break down.)* Explain that decomposition is an important process because, without it, there would be no nutrients in the soil for the next generation of plants and animals to use. Soil houses many different types of organisms that help break down organic materials. Most of the time you don't see decomposition at work in soil because it is hidden from view.

❺ Hand out copies of the "Let It Rot" lab sheet and invite students to investigate what factors aid in the decomposition process.

Let It Rot

What conditions promote decomposition in nature?

❶ Fill each of the two short soda bottle cylinders with a mixture of leaves, grass clippings, vegetable scraps, and shredded newspaper, up to about 2½ centimeters (1 in.) from the top. Spray the mixture in one bottle with water until it is very damp. Leave the other mixture dry.

❷ Take one of the taller soda bottle cylinders and turn it upside down over one of the smaller ones. Slip the tapered end of the taller cylinder inside the shorter one and push them together so they fit snugly together. Repeat the same procedure with the second cylinder. Why do you think the two cylinders must fit together tightly?

You'll Need

- 2 clear 2-liter soda bottles with top cut off 10 cm from the bottom
- 2 clear 2-liter soda bottles with top cut off 23 cm from the bottom
- thumbtack or pushpin
- shredded newspaper, old leaves, grass clippings, and/or vegetable scraps
- spray bottle or plant mister
- cellophane tape

_____

_____

❸ Using a pushpin or thumbtack, punch about 15 small holes into the top cylinder of the compost chamber with the wet mixture. Do not put any holes in the other compost chamber–instead, use tape to seal the connection where the two cylinders join together.

❹ Based on how you constructed the two compost chambers, what are the two main environmental differences between the two organic mixtures?

_____

❺ Predict: What changes do you expect to see in each bottle over time? Write your predictions here:

_____

_____

❻ Carefully observe each bottle for the next seven days and record your observations on the chart below. If the bottle with the holes begins to dry out, take it apart and spray more water in it.

Day	Wet Bottle	Dry Bottle
1		

(continued)

Day	Wet Bottle	Dry Bottle
2		
3		
4		
5		
6		
7		

❼ Based on your observations, under what conditions does natural decomposition work best?

_____

_____

Think About It: What other environmental factors might affect the speed of decomposition? How might you test them?

Soda Bottle Science Scholastic Teaching Resources